The Latin American Studies Book Series

Series Editors

Eustógio W. Correia Dantas, Departamento de Geografia, Centro de Ciências, Universidade Federal do Ceará, Fortaleza, Ceará, Brazil

Jorge Rabassa, Laboratorio de Geomorfología y Cuaternario, CADIC-CONICET, Ushuaia, Tierra del Fuego, Argentina

Andrew Sluyter, Louisiana State University, Baton Rouge, LA, USA

The Latin American Studies Book Series promotes quality scientific research focusing on Latin American countries. The series accepts disciplinary and interdisciplinary titles related to geographical, environmental, cultural, economic, political and urban research dedicated to Latin America. The series publishes comprehensive monographs, edited volumes and textbooks refereed by a region or country expert specialized in Latin American studies.

The series aims to raise the profile of Latin American studies, showcasing important works developed focusing on the region. It is aimed at researchers, students, and everyone interested in Latin American topics.

Submit a proposal: Proposals for the series will be considered by the Series Advisory Board. A book proposal form can be obtained from the Publisher, Juliana Pitanguy (juliana.pitanguy@springer.com).

More information about this series at http://www.springer.com/series/15104

The Latin American Studies Book Series

Series Editors

Eustógio W. Correia Dantas, Departamento de Geografia, Centro de Ciências, Universidade Federal do Ceará, Fortaleza, Ceará, Brazil

Jorge Rabassa, Laboratorio de Geomorfología y Cuaternario, CADIC-CONICET, Ushuaia, Tierra del Fuego, Argentina

Andrew Sluyter, Louisiana State University, Baton Rouge, LA, USA

The Latin American Studies Book Series promotes quality scientific research focusing on Latin American countries. The series accepts disciplinary and interdisciplinary titles related to geographical, environmental, cultural, economic, political and urban research dedicated to Latin America. The series publishes comprehensive monographs, edited volumes and textbooks refereed by a region or country expert specialized in Latin American studies.

The series aims to raise the profile of Latin American studies, showcasing important works developed focusing on the region. It is aimed at researchers, students, and everyone interested in Latin American topics.

Submit a proposal: Proposals for the series will be considered by the Series Advisory Board. A book proposal form can be obtained from the Publisher, Juliana Pitanguy (juliana.pitanguy@springer.com).

More information about this series at http://www.springer.com/series/15104

Luiz Cesar de Queiroz Ribeiro · Filippo Bignami
Editors

The Legacy of Mega Events

Urban Transformations and Citizenship in Rio de Janeiro

With Contributions by Ana Paula Soares Carvalho, Humberto Meza, Niccoló Cuppini, Orlando Alves dos Santos Junior

Editors
Luiz Cesar de Queiroz Ribeiro
INCT Observatório das Metrópoles
Rio de Janeiro, Brazil

Filippo Bignami
University of Applied Sciences and Arts of
Southern Switzerland, SUPSI
Lugano, Switzerland

ISSN 2366-3421 ISSN 2366-343X (electronic)
The Latin American Studies Book Series
ISBN 978-3-030-55055-4 ISBN 978-3-030-55053-0 (eBook)
https://doi.org/10.1007/978-3-030-55053-0

This Springer imprint is published by the registered company Springer Nature Switzerland AG
The registered company address is: Gewerbestrasse 11, 6330 Cham, Switzerland

Acknowledgement

Prof. Luiz Cesar de Queiroz Ribeiro and Dr. Filippo Bignami as General Editors thank Ana Paula Soares Carvalho, Humberto Meza, Niccoló Cuppini and Orlando Alves dos Santos Junior for their contributions as co-editors of this book.

About This Book

This book is originated from the action research project titled *Urban Regimes and Citizenship: A Case Study for an Innovative Approach* carried out by the National Institute of Technology and Science (INCT)/Observatório das Metrópoles, headquartered in the Urban and Regional Planning and Research Institute (IPPUR) of the Federal University of Rio de Janeiro (UFRJ) and the University of Applied Sciences of Southern Switzerland (SUPSI), Department of Economics, Health and Social Sciences (DEASS). Pontifícia Universidade Catolica do Rio de Janeiro (PUC-Rio) as well actively cooperated in the activities.

The project, started in the end of 2017 with a 2-year duration, is financed by the Swiss National Science Foundation (SNSF) on the Swiss side and by Fundação Carlos Chagas Filho de Amparo à Pesquisa do Estado do Rio de Janeiro (Faperj) (Carlos Chagas Filho Foundation for Research Support of the State of Rio de Janeiro) and Conselho Nacional de Desenvolvimento Científico e Tecnológico (CNPq) (Brazilian National Council for Scientific and Technological Development) on the Brazilian side.

The project accomplished its activities with a methodology of action research, allowing a fruitful combination of outcomes, encompassing scientific articles, a module of citizenship education tested in a Rio de Janeiro secondary school, quantitative and qualitative surveys and a video firstly having an educational scope, but as well popular, social and political purposes, available at www.youtube.com/watch?v=l5qiR2bwXl0&t=9s.

Introduction

Rio de Janeiro was the scenario where in the decade 2006–2016 a wave of mega events (FIFA World Cup, Olympic Games, Young Olympics, and others) took place. Now, these glittering events are over, and the big, unexplored, question is: what is happening to the metropolis in consequence of such period? Or in other words, what is the legacy of these mega events for the city?

Much attention has been devoted to the process of planning, hosting, and organizing these events. Studies and analyses exist on the *pre*-mega events,[1] but the *post*-mega events have not yet been probed exhaustively. A multidisciplinary glance at the legacy of the wave of a decade of mega events is the gap that this book aims to bridge in the case of Rio de Janeiro. This collective work, being both an outcome of the above-described project and a wider reflection done by academics in a multidisciplinary approach, aims to identify the key drivers of urban evolution in relation to the mega events hosted in Rio in the last decade, and to enlighten the framework and the impact that emerged after them. The main theories adopted in the project originating this book are urban regimes (Savitch and Kantor 2002) and citizenship (Isin 2008; Sassen 2005; Walzer 1989; Dagger 1997; Rawls 2005), aiming to deepen the evolution of the urban transformation in the city featuring the decade of mega events, linked with key aspects of citizenship, such as social and political conditions hampering or enabling processes of inclusion and exclusion, with a specific glance at the role of citizenship education.

The analysis of transformations in urban governance is part of the debate on urban regimes, insofar as they allow us to advance the understanding of the conditions of governance and of the institutional capacities of the municipality of Rio de Janeiro, opening the way for identifying the challenges or obstacles to urban development. The urban regime approach has gained recognition in the Anglo-Saxon literature on local politics by refuting both structuralist approaches

---

[1]To understand the debate around the pre-mega events, we suggest the Dossier of the Popular Committee of the World Cup and Olympics in Rio de Janeiro retrieved from: http://issuu.com/mantelli/docs/dossiecomiterio2015_issuu_01, as well Santos Junior, O. A., & Lima, C. G. R. (2015).

and the methodological individualism of public choice theory, revaluing political dynamics in urban political economy analyses (Harding 1999). It can be stated, in a succinct and direct way, that urban systems theory intends to investigate how cities are ruled. In this sense, it is important to understand the context of bargaining, as proposed by Kantor and Savitch (1997), according to which particular bargaining environments affect urban regime structure by highlighting the importance of the political and economic context that influence such environments.

This book, collecting ten contributions on the topic, wants then to grasp the manifold aspects of what the mega events entailed for Rio de Janeiro, positioning the city in a frame of urban transformation as configuration of global city. The process of neoliberalization and its key actors influencing the decision-making in a frame of urban regime approach (Brenner et al. 2011; Peck 2015), the effects and the projection of the urban transformation on the social and political environment and the nexus between the city and the multifaceted concept of citizenship (Isin 2007; Sassen 2005; Nyers and Rygiel 2012; Holston and Appadurai 1999) including the following.

The Neoliberal Urban Renewal of the City of Rio de Janeiro

Through investments made in systems of urban mobility, expressways, viaducts, tunnels and infrastructure networks, it is possible to apprehend that a set of deep urban transformations is underway in Barra da Tijuca, in the Port Area, and in the South Zone of the city of Rio de Janeiro. The majority of these investments were funded by resources associated with the preparation of the city to host the mega-sporting events, especially the 2016 Summer Olympics.

Although there is investment in other areas of the city, the hypothesis that can be drawn from the studies gathered in this book is that these three areas—Barra da Tijuca, Port Area and South Zone—are experiencing such neoliberal urbanization processes (Theodore et al. 2009), which makes it necessary to evaluate the trans-formations that have come through.

In this sense, it seems interesting to start from the idea of spatial fix, as proposed by Harvey (2005; 2008; 2012), in the city of Rio de Janeiro. In fact, a kind of neoliberal spatial fix. The older urban setting, among other characteristics, seemed to be marked by the abandonment and devaluation of the central area; the complex relationship between physical proximity and social distancing in the coexistence of favelas with the south zone of the city and the significant devaluation of the environment of the favelas in these areas; the housing boom, directed to the middle and upper classes toward the most prized area west of the city, Barra da Tijuca and Recreio dos Bandeirantes; and finally, the process of heterogeneity of the metropolitan periphery, with the emergence of residential areas of well-structured middle-class spaces, accompanied by the continued expansion of favelas and pre-carious neighborhoods.

It seems possible to identify changes in this spatiality that moves toward four urban settings: (i) the intensification of the appreciation and elitization of Barra da Tijuca, which progressively becomes an area not merely of real estate expansion, but an area of services and businesses; (ii) the appreciation of the central area with a view to attract residences aimed for the middle and upper classes; (iii) the enhancement of the environment of the South Zone favelas themselves, particularly their areas privileged by their location, which would attract a segment of the middle classes; and (iv) the continuing expansion into the metropolitan periphery, both in the perspective of the growth of favelas and of the diversification of residential cores geared toward middle and upper income residents.

What is important to point out is that these changes in spatiality and the emergence of this complex urban setting could not be a result of randomness but would be the local expression of neoliberal spatial adjustment promoted by the neoliberal entrepreneurial governance that, in a different fashion, impacts the cities of core countries (Hackworth 2007).

In the case of Rio de Janeiro, we can see the active role of government in promoting the observed changes, not limited to enabling the urban renewal projects to be conducted by private capital. In this perspective, Rio de Janeiro City Hall appears as the main promoter of urban renewal projects being implemented, acting in different ways, involving the coordination or preparation of projects, the direct financing of various interventions, the granting of tax incentives and tax exemptions for the attraction of private companies, the adoption of new institutional arrangements for management of urban space and changes in previously existing legislation, particularly those related to construction parameters.

In this process, one cannot fail to mention the participation of other government spheres, both federal and state government administrations, particularly with regard to direct investments and financing of interventions, such as the works of mobility of the Bus Rapid Transit (BRT), the Light Rail System (VLT in Portuguese acronym) and the subway. Also, the transformations in urban settings linked to Barra da Tijuca, Port Area, and South Zone (see the Map in Fig. 1.4 in Chap. 1) would be associated with real estate valuation processes, gentrification, and social elitization.

The chapters in this book make it possible to interpret the current processes of urban transformation of the city of Rio de Janeiro as a process of neoliberalization expressed in a new round of commodification, associated with a process of creative destruction involving urban configurations, institutional arrangements, and urbanistic and social regulations involving these areas of the city. Despite the focus on certain areas of the city (mostly Barra da Tijuca), these urban transformations deepened the abandonment and segregation process on precarious regions such as the North Zone neighborhoods and the Metropolitan Region municipalities, specifically inside the Baixada Fluminense sub-region. It is important to raise some impacts of such neoliberal modernization upon urban governance and the future of the city of Rio de Janeiro.

Firstly, based on the approach specified here, it is worth highlighting that such process of changes preserves old agents, practices, urban structures, institutions, as well as institutional arrangements, and it occurs combined with what was preserved. Therefore, the attempt here was to show that what is going on is not merely a continuity of previous processes. There are indeed new processes that do not exactly express a rupture with old practices, but an inflection, in which the ongoing neoliberal modernization can be considered conservative in several aspects. Anyway, it is possible to infer that from the point of view of urban governance, this neoliberal modernization seems to approach the patrimonial practices that both mark the history of the city of Rio de Janeiro and distance themselves from the democratic management associated with the ideas of the right to the city. In this context, the public spheres of participation are progressively replaced by decision-making processes that subordinate the government to market logics.

Secondly, in line with the approach used in this intro (Brenner et al. 2011), the implementation process of this neoliberalization project involves several contradictions and raises different urban conflicts comprising resistance and opposition, for example, with regard to the priorities of investments, the removals of communities located in the areas of intervention, the lack of social participation channels, and changes in social life. Such conflicts, perpetrated by a variety of organizations and social movements, may relate to the neoliberalization project course, changing more or less substantially, or even invalidating it, depending on the strength that it will achieve over time, which reinforces the uncertainty about the future of the city. In this sense, one can predict that the urban governance of Rio de Janeiro tends to be marked by the intensification of conflict.

However, as a third aspect, one cannot ignore the power of the private–public coalition that commands this neoliberal entrepreneurial governance project, which demonstrates hegemonic strength and ability to incorporate, in a subordinate manner, at least discursively, subaltern interests. In this way, it complements the action of other agents and existing political grammars to enable its implementation, resulting in the specificity of the neoliberal city of Rio de Janeiro. In short, the profound transformations in the urban dynamics of Rio de Janeiro comprise, on the one hand, new commodification processes of the city and, on the other, new patterns of relationship between the government and the private sector, characterized by the government's subordination to the private actors, especially the real estate sector. This process involves the creative destruction of physical structures, institutional arrangements, and urban and social regulations that would aim to create new conditions for the production and reproduction of capital in the context of contemporary globalization, expressing a new structured coherence (Harvey 2005), which simultaneously preserves urban structures, social institutions, and agents in the territory. The combination of old and new moves toward the reproduction of practices threatens the principles of democratic governance and the universalization of the right to the city.

The purpose of this book is to inquire the manifold aspects of what the mega events entail for Rio de Janeiro, framing the city in an urban transformation as configuration of global city. It contributes to understand the context and conditions in which coalitions of power emerged and acted in the city, expressed in an urban regime approach, with a specific perspective on citizenship implications.

The Feature of the Mega events' Cycle and the Failure of the Process

The mega events' cycle expressed a stronger futuristic and radiant fantasy of the city's cosmopolitan elites. At the beginning of the twentieth century, the radical reforms of the city promoted by the former mayor Pereira Passos promised to transform the decadent and unhealthy center of the city into a "Paris in the Tropics", capable of rivaling the enviable Buenos Aires and its *belle époque* urbanism (Abreu 2010). In the 1970s, the conservative modernism of the developmental elites opened the city's boundaries to the west, incorporating the area known as Barra da Tijuca and its 165.59 km by building a system of viaducts and tunnels. It was now the futuristic promise of turning Rio de Janeiro into a "Miami in the Tropics" (Ribeiro et al. 2017). The preparation of the city to host the mega events was part of a new cycle of fantasies of the city's cosmopolitan elites. It is now to make Rio de Janeiro a "Global and Creative City", a successful example of urban entrepreneurship capable of reversing its long trajectory of crisis and stagnation, turning it into a global and radiant city. Pushed and promoted from the local elites and the local government (implying then a political and economic improvement), a climate of optimism and redemption was spreading in society based on an epic narrative actively diffused by the media powers, having as object the mythical promise of the "legacies" of the mega events. The belief in this narrative was fueled by the social perception of bonanza experienced by the population with the expansion of employment, real growth in labor income, social inclusion by the consumption of the vast segment of the poor, decrease in violence rates, etc. In the collective imagination and in the social experience, it seemed to everyone that the city of Rio de Janeiro would return to its destination as the "Wonderful City".

This last fantasy cycle of elites failed, as well as the other two mentioned above. Even before the completion of the last mega event (the Olympic Games), one could see the embedded signs of the economic, social, urban, fiscal, and institutional crisis in which the city (and the State of Rio de Janeiro) had already been immersed. The so-called legacies have become new (and serious) problems to be solved. This is the case of the crisis of urban mobility in which the city is plunged, with the physical and institutional decomposition of the BRT model or the fiscal crisis of the public sector, as an example. The promise of the pacification of the city turned into an explosion of volition, an increase in homicide rates and a resumption of a more virulent security policy based on the logic of the wrath waged in the popular territories.[2]

---

[2]The evidence of this set of crises is abundant. In 2018, the public transport sector, controlled by FETRANSPOR (a network of private companies running public transportation) registered a fall in demand by 15–20%, while the largest company in the network was involved in a corruption scandal leading to the arrest of its president. In terms of public security, data from the former Public Security Institute (ISP) of the State of Rio de Janeiro recorded that the police murder rates for 2017 (1,124 cases) approached data for 2007 (1,330 cases), all of which are associated with a restrictive scenario characterized by the austerity in the official discourse for the public investments, product of an annual deficit of approximately R $ 16 billion, putting Rio de Janeiro in a state of calamity.

Actually, there is much evidence that the cycle of mega-sporting events and its promises of redemption is a failed experiment from various points of view. How can one understand this new failure of the cosmopolitan elites and their futuristic and redemptive ambitions? (Molina Silveira 2013). There are many possible causes, but it is not for this introduction to explore this theme. But it certainly stems from the absence of a long-term city development view on the entire urban project in which the cycle of urban renewal could have functioned as a powerful lever, pushing ahead the positive impulses generated by public and private investment. The absence of such a development view within the municipal project stems from the nature of the political economy that historically governs the cycles of urban reengineering in which there remains a renewed coalition of interests anchored in urban accumulation. In other words, as already demonstrated in another work (Ribeiro et al. 2017), it has now (as in the other two cycles) the effects of path dependency that subordinate the trajectory of the city to interests that maintain the city as a growth machine much like the one described by Logan and Molotch (1996).

City and Citizenship: A Dyad in Tension in Rio de Janeiro

The metropolis of Rio de Janeiro was a formidable setting of a process of urban transformation having mega events as engine, featured by a substantial failure at least in terms of shared benefits for the population, not only in the economic aspect, but also in the failure to strengthen the social and political scenarios. As inferable from the two above sections, besides sound gains for a restricted *élite*, this process generated a wide relapse on the collectivity, especially in some areas through extended gentrification plans (Gaffney 2015), insofar worsening the democratic life of the city and further hampering the possibility to feed a virtuous relation of participation, of a political constructive debate between its citizens. Thus, it appears to be a clear separation between the provisions for the city and the possibility to create a course of entitlement for its collectivity. This separation is surely a meaningful issue in a metropolis such as Rio de Janeiro, already affected by profound social and political knots, but has the value to pose the question of what urban citizenship means in terms of *entitlement* and *provision* (Dahrendorf 2007) to enable a substantial citizenship: the socio-political debate must be able to include citizens of a city with the means to reach the goal of deliberating on the changes to be implemented in the city itself (in its structure, its polity, its policies, its social settlement). The intimate connection between city and citizenship is then fed intending the right to participate and have a say not as a mere right, but as an explanatory concept (a tool, we could say) which re-defines what is possible to perform in a participatory approach in the course of pursuing citizenship rights. This connection is exactly what the process briefly introduced above further spoiled in this metropolis.

This represents a profound contrast, considering the nature of this link between city and citizenship. City derives in effect from the Latin *civitas*, meaning the community of its dwellers, *cives*, and implies the acknowledgement of being a citizen on social, political, economic, and juridical aspects. Another Latin word for city is *urbs*, meaning essentially the physical space of buildings, an area with a precise perimeter, with an "inside" and an "outside" (Pocock 1998). With the link city-*civitas*, the emphasis is on the inclusive perspective, indicating the collectivity of citizens living in a shared, co-constructed, and co-ruled space, since "(…) *the city is the product of the common action or interaction of the citizens, and not the reverse*" (Balibar 2008, p. 523). In this way, we have simply depicted the three pillars constituting the existence of a city: a collectivity, intended as social and political aggregate; a shared and co-constructed political process achieving a set of rules; a space to be planned and used appropriately. We are then fully in the *fulcrum* of the discussion about how the process of urban transformation is impacting Rio de Janeiro as a political and social urban fabric.

The city is the most significant material and immaterial space where the dimensions of citizenship are shaped and concretized (Isin 2007; Sassen 2005; Nyers and Rygiel 2012). It enables right and duties, participation, identity, and membership. Further, citizenship ensures the uniformity of rights and duties linked with political participation, and therefore has the potentiality to mitigate political effects of social inequalities. Citizenship has been discussed under the perspective of an urban setting as a possible contemporary alternative to long-established notions of citizenship, those built on the pillars of rights, duties, identity, and membership to a political entity, usually a nation-state (Purcell 2003). The urban citizenship standpoint started to be discussed more vigorously at the end of the twentieth century, when it was clear that the nation-state was not adequate anymore to collocate with the concept of citizenship (Sassen 2002). In reaction to diverse pressures, simplistically definable using terms such as neoliberalism, globalization, social inequality, economic rescaling, migration flows, spatial segregation, invasive corporate control, and *financial pantheism* (Bignami 2017, p. 133), urban citizenship has emerged to better reflect the identity, participation, and entitlement to exploit the city from the city dwellers themselves.

This book, by delving into the process of urban transformation, also sheds light on the link between city and citizenship, evidencing that the process of the mega events' implementation in Rio is producing an increased difficulty and fences in a diffused political action achieving a *performative citizenship* (Isin 2017) at urban level.

The application of such a perspective allows a profound reflection firstly on how to equip citizens with the capacity to be aware of their potentiality. In a nutshell, it is crucial understanding of how to increase the competence of the social, political, and economic (potential) power the concept of citizenship implies. It is crucial shedding light on this topic and furthermore to act to pursue this aim. Indeed, referring to the citizenship issue in the Rio de Janeiro frame, this book wants to open up the reflection upon the necessity to determine: (a) the extent of the "often" permissibility of disruption and contention, especially in the presence of huge urban transformations generated by a series of mega events; (b) the ways to co-construct a political process implying a participatory process, necessarily through a new conception of citizenship education; and (c) the shared standpoint that being an inhabitant is neither a strict condition, nor the only condition; being a performative citizen, in the sense of a member of a socio-political community acting to shape its material and immaterial architecture, is the single sufficient precondition.

Such set of entitlements is to inform not only legal codes but also predispositions and understandings of the actors involved in the public sphere that can act as catalysts for collective efforts toward social justice and for a disjuncture in the re-set of political and social actions able to "fill" the city with citizenship.

Book Structure and the Chapters

The book has ten chapters discussing many aspects of the urban transformations of Rio de Janeiro Metropolitan Area after a decade of mega events and shows social resistances and several approaches around the urban transformation that affected the metropolis and its social, political, and economic implications. The ten chapters then frame Rio de Janeiro at the light of the mega events and, in doing so on a common basis, ideally handle three observation perspectives: the first is the description of urban transformations and mega events (chapters by de Queiroz Ribeiro and dos Santos Junior; Cuppini; dos Santos; Sampaio); the second addresses neighborhoods as case studies representing an ensign of neoliberal urban as case studies representing an ensign of a neoliberal urban transformation' results (chapters from Borba; Meza; dos Santos Junior, Lacerda, Werneck and Ribeiro); the third links city and citizenship focusing tensions and inconsistencies and opening up a perspective on the importance of fostering the concept of citizenship (chapters from Novaes; de Castro; Bignami and Soares Carvalho).

This ideal tripartition includes the awareness of the debate that emerged with Rio de Janeiro's graduation as a "global city" (Sassen 1991). Our chapters adopt a multidisciplinary approach to analyze both the legacies of mega events and the transformations that came after this global wave. The adoption of the "global city" category requires the observation of the institutional convergences (of local, state, and federal interfaces), as well as the interactions of the various agents (state, private capital, multinational corporations, and social movements) in the maintenance of the current urban order. Therefore, the book has a sound potential to understand, on one hand, the legacy of mega events in terms of the actual setback and, on the other hand, to trace a precise map of key features and concepts in order to drive a perspectival and updated understanding of the subject.

Hence, the chapters create a dialogue and a logic sequence throughout the three observation perspectives, where the reader can firstly deepen the concepts featuring the process of transformation (such as neoliberal urbanization process, globalization, and failure) and gain awareness on how these concepts are exploited and reified within the metropolis. Then, scaling down on study cases in terms of placing the lens on neighborhoods (in particular, Porto Maravilha but also beyond), the reader can find the clue to concretely place such lens in the metropolis where the state has abdicated the control role over the private agents, assuming the task of enabler of the process. Finally, citizenship issue is claimed as an intricate tie with the city, on one hand in terms of how the process of urban transformation creates gaps, hampering democratic conditions and preventing participatory and performative acts. On the other hand, citizenship is framed in theoretical terms in the Rio de Janeiro setting, speculating on its educational aspect.

The book has the following chapters' structure.

Chapter 1, written by Luiz Cesar de Queiroz Ribeiro and Orlando Alves dos Santos Junior, discusses how the current developments in Rio de Janeiro in the wake of the 2014 FIFA World Cup and the 2016 Summer Olympics have accelerated the transition from a hybrid urban order to a new urban order of intensified neoliberalization and commodification. As the authors argue, this transition is a complex and contradictory process that entails both roll-back and roll-out features of neoliberalism, as well as a permanent restructuring of socio-spatial relations and political arrangements. The chapter examines the creative destruction of urban structures, institutional arrangements, and regulations of the urban space of three sites in Rio de Janeiro—Barra da Tijuca, the Port Area, and the South Zone. In short, the authors underline that such destruction can be understood as a process of variegated neoliberalization.

Chapter 2 is written by Niccolò Cuppini. It presents a historical background of Rio de Janeiro development, discussing the project of transforming it into a global city through the analysis of its globalizing strategies and the implementation of mega events. The author presents an analysis of the legacy of these processes, mainly focusing on the question of the transportation system and public security.

In Chap. 3, Carolina dos Santos analyzes one of the most important tools (PPPs) used by the Rio Municipality for most of the urban operations in the city around the Olympic Games. Dos Santos verified how the discourse of city improvements, triggering the idea of a social legacy, projects of renovations, and urban restructuring benefited and legitimized the speculative capital without a broad social debate and not taking into account the interests of the population, especially the popular classes.

Fernanda Sampaio introduces an interesting point of view to analyze one of the most important cultural traditions of Rio: Carnival. Her Chap. 4 points out the disputes around the control for this cultural patrimony, specifically the "street Carnival", which from that moment on began to acquire the structural and institutional mega event features, leading, in addition to the "mercantile treatment of the party", to commodification and privatization of the city's public spaces. The mechanisms adopted by the public sector for the organization and realization of the carnival followed the same format of other urban intervention projects for this new city model.

Tuanni Borba, author of Chap. 5, analyzes the institutional trajectory of the SPU (one of the institutions of the federal government responsible for the national cultural and material patrimony) and how it was cultivated, based on the performance of its superintendence in Rio de Janeiro (SPU-RJ) within the Porto Maravilha project, the largest urban operation in Brazil. From the perspective that change depends on investment by the endogenous actors, the bifurcation illustrates the fact that these investments, explained in terms of cultivation or neglect, are not exclusive because the same actors cultivate or neglect certain trajectories depending on the interests at stake. This is the main hypothesis of the chapter: the coexistence of institutional trajectories in the SPU-RJ.

In some way, the sequence of this debate is present in the chapter written by Humberto Meza (Chap. 6). He analyzes the first serious crisis experience by the huge urban operation in the middle of 2017 as a consequence of the financial and political crisis that affected Brazil after the impeachment of Dilma Roussef and the corruption investigation of the Operation "Car Wash" (*OperaçãoLava Jato*). His main hypothesis is that the crisis of 2017 brought to the fore a diversity of actors, in addition to the large contractors and state actors that make up the PPP. This *actor's moving* reveals, at the same time, different roles of the Brazilian State to guarantee the private participation in the urban operation.

Chapter 7, written by Orlando Alves dos Santos Junior, Larissa Lacerda, Mariana Werneck, and Bruna Ribeiro, aims to discuss the first results of the survey developed in the port area of Rio de Janeiro around the existing slum tenements (called *cortiços*, which are several buildings with shared bedrooms where many lower class families live together) in the region, their housing conditions, and the profile of their population. They sought to deconstruct the current perception that stigmatizes these spaces—and their inhabitants—as precarious and marginal, showing that the slum tenements are marked by a great heterogeneity of housing conditions and social groups, unified in their demand to live in the central area of the city.

For the following debate, Patricia Novaes analyzes the favelas phenomena from an innovative perspective for the scholar field. In her chapter (Chap. 8), she delivers a reflection around the emergence of new ventures aimed at a middle-class audience within favelas in Rio, most of the South Zone. This process represents a sort of reversal of public investments in the favelas creating, at the same time, the reversal of the symbolic resignification and gentrification experiments process inside the popular territories as consequence of the mega-sporting events with a steadily increasing tension between city and citizenship.

In Chap. 9, Taiana de Castro Sobrinho continues the analysis on favelas and proves the hypothesis that the removal cycle in Rio around the global events was the result of a pro-market-oriented urban policy based on a strategic planning that contributed to increase the vulnerability of dwellings in favelas and to the invisibility of issues related to the protection of these dwellings. The author remarks on the importance of pointing out that no mechanisms were created to avoid the expulsion of the poor from the areas of economic interest of the city and its transfer to remote and peripheral areas, which contributed to highlight the socio-spatial segregation in the urban space. This chapter describes then a concrete perspective of citizenship tensions reified in the city.

The Chap. 10, written by Filippo Bignami and Ana Paula Soares Carvalho, presents in a first part a theoretical discussion on the potential contributions of citizenship education for the development of the students' sense of political and social membership. As the authors argue, practicing citizenship education is fundamental to build a more peaceful, just, and democratic city in which to live. In the second part of their chapter, the authors delve into action, resuming an innovative module of citizenship education set up and successfully tested in a secondary school of Rio de Janeiro within the bilateral project carried out.

Luiz Cesar de Queiroz Ribeiro
Institute of Urban and Regional Planning
Federal University of Rio de Janeiro
Rio de Janeiro, Brazil

Filippo Bignami
University of Applied Sciences and Arts of Southern Switzerland (SUPSI)
Lugano, Switzerland
e-mail: filippo.bignami@supsi.ch

Orlando Alves dos Santos Junior
Institute of Urban and Regional Planning
Federal University of Rio de Janeiro
Rio de Janeiro, Brazil

References

Abreu MA (2010) Evolução Urbana do Rio de Janeiro. 4. ed. Rio de Janeiro: Instituto Pereira Passos

Balibar E (2008) Historical dilemmas of democracy and their contemporary relevance for citizenship. Rethink Marx 20(4):522–538. doi: 10.1080/08935690802299363

Bignami F (2017) Going intercultural as a generative framework of a respondent citizenship. In: Onorati MG, Bignami F, Bednarz F (eds), Intercultural Praxis for Ethical Action. Reflexive Education and Participatory Citizenship for a Respondent Sociality. EME publications, Louvain, Belgium

Brenner N, Peck J, Theodore N (2011) Y despés de la neoliberalización de las transformaciones regulatorias contemporâneas. URBAN. Revista del Departamento de Urbanistica y Ordenacion del Territorio. Marzo, 2011

Comitê Popular da Copa e Olimpíadas Do Rio de Janeiro (2015) Megaeventos e violações dos direitos humanos no Rio de Janeiro—Dossiê do Comitê Popular da Copa e Olimpíadas do Rio de Janeiro. Olimpíada Rio 2016, os jogos da exclusão 2015 [Mega-events and human rights violations in Rio de Janeiro—Dossier of the Popular Committee of the World Cup and Olympics in Rio de Janeiro. Rio 2016 Olympics, the 2015 exclusion games]. Retrieved from http://issuu.com/mantelli/docs/dossiecomiterio2015_issuu_01 Dagger, R. (1997). Civic Virtues. Rights, Citizenship, and Republican Liberalism. Oxford: Oxford University Press

Dahrendorf R (2007) The Crisis of Democracy.Gibson Square Books, London

Gaffney C (2015) Gentrifications in pre-Olympic Rio de Janeiro. Urban Geogr 1132–1153 doi: 10.1080/02723638.2015.1096115

Hackworth J (2007) The neoliberal city: Governance, ideology, and development in American urbanism. Cornell University Press, New York, NY

Harding A (1999) Review Article: North Urban Political Economy, Urban Theory and British Research, Br J Polit Sci 29(4):673–98

Harvey D (2005) A produção capitalista do espaço [The capitalist production of space]. Annablume, São Paulo, Brazil

Harvey D (2008) O neoliberalismo: história e implicações [Neoliberalism: history and implications]. Loyola, São Paulo, Brazil

Harvey D (2012) Rebel cities. London, England: Verso Holston J, Appadurai A (1999). Cities and Citizenship. In: Holston J, Appadurai A (eds), Cities and Citizenship (pp 1–18). Durham, NC: Duke University Press

Isin EF (2007) City.State: Critique of Scalar Thought. Citizh Stud 11(2):211–228. doi: 10.1080/13621020701262644

Isin EF (2008) Theorizing acts of citizenship. In: Isin EF, Nielsen GM (eds), Acts of Citizenship. Palgrave Macmillan, London, pp. 15–43

Isin Engin F (2017) Performative Citizenship. In: A. Shachar, S., Bauböck, R., Bloemraad, I., Vink, M. (Ed.), The Oxford Handbook of Citizenship. Oxford University Press, Oxford, pp 500–523

Logan JR, Molotch HL (1996) The City as a Growth Machine in Fanstain, Susan; Campbell, Scott (Eds.) Reading in Urban Theory. Blackwell Publishers, Massachussetts

Molina FS (2013) Mega-eventos e produção do espaço urbano no Rio de Janeiro: da "Paris dos Trópicos" à "Cidade Olímpica". Doctoral Dissertation. São Paulo: Faculdade de Filosofia, Letras e Ciências Humanas da Universidade de São Paulo/Departamento de Geografia

Nyers P, Rygiel K (eds) (2012). Citizenship, Migrant Activism, and the Politics of Movement. Routledge, New York

Peck J, Theodore N (2015) Fast policy experimental statecraft at the thresholds of neoliberalism. University of Minnesota Press, Minneapolis

Pocock JGA (1998) The ideal of citizenship since classical times. In: Shafir G (ed), The citizenship debates: a reader. University of Minnesota Press, Minneapolis

Purcell M (2003) Citizenship and the Right to the Global City: Reimagining the Capitalist World Order, Int J Urban Reg Res 27(3):564–590

Rawls J (2005) Political Liberalism. Expanded edition. Columbia University Press, New York

Ribeiro LCQ (2017) Metamorphose of the Urban Order of the Brazilian Metropolis: The case of Rio de Janeiro, In: Ribeiroe et al (ed) Urban Transformations in Rio de Janeiro: Development, Segregation, and Governance. Springer Publishing, 2017

Santos Junior OA, Lima CGR (2015) Impactos econômicos dos megaeventos no Brasil: Investimento público, participação privada e difusão do empreendedorismo urbano neoliberal [Economic impacts of mega events in Brazil: Public investment, private participation and diffusion of neoliberal urban entrepreneurship]. In: Santos Junior OA, Gaffney C, Ribeiro LCQ (eds), Brasil: Os impactos da Copa do Mundo 2014 e das Olimpíadas 2016

Sassen S (1991) The global city. New York, London. Tokyo, Princeton: Princeton University Press

Sassen S (2002) The Repositioning of Citizenship: Emergent Subjects and Spaces for Politics. Berkeley J Sociol 46

Sassen S (2005) The Repositioning of Citizenship and Alienage: Emergent Subjects and Spaces for Politics. Globalizations 2:79–94

Savitch HV, Kantor P (2002) Cities in the international marketplace. The political economy of urban development in North America and Western Europe. Princeton University Press, Princeton and Oxford

Theodore N, Peck J, Brenner N (2009) Urbanismo neoliberal: la ciudad y el imperio de los mercados. Temas sociales, Santiago de Chile, n. 66, marzo

Walzer M (1989) Citizenship. In: Ball T, Farr J, Hanson RC (eds), Political Innovation and Conceptual Change. Cambridge University Press, Cambridge

By the Brazilian Side:

Research Coordination

OBSERVATÓRIO
DAS METRÓPOLES
Instituto Nacional de Ciência e Tecnologia

IPPUR

Instituto de Pesquisa
e Planejamento Urbano e Regional

UFRJ

Supported by the Financial Agencies

By the Swiss Side:

Research Coordination

University of Applied Sciences and Arts of Southern Switzerland
Department of Business Economics, Health and Social Care

SUPSI

Supported by the Financial Agency

SWISS NATIONAL SCIENCE FOUNDATION

Contents

Editors and Contributors

About the Editors

Luiz Cesar de Queiroz Ribeiro is Professor at the Federal University of Rio de Janeiro—Urban and Regional Planning and Research Institute—IPPUR/UFRJ. He holds a Ph.D. in Architecture and Urbanism by the University of São Paulo, USP and Master in Economic and Social Development by the Université Paris 1 Pantheon-Sorbonne, Coordinates the National Institute of Science and Technology, INCT *Observatório das Metrópoles*: a research network involving comparative studies in 15 Brazilian metropolises around territory, social cohesion, and governance issues. His research areas are metropolitan, intrametropolitan dynamics and the national territory, socio-spatial dimension of the exclusion/Integration in the metropolis, Urban governance, citizenship, and management of the metropolis. He is Editor of Cadernos Metrópoles and e-metropolis magazines.

Filippo Bignami holds a Ph.D. in Political and Social Sciences. He is Senior Researcher and Lecturer at the University of Applied Sciences of Southern Switzerland, Department of Economics, Health and Social Sciences (SUPSI-DEASS) —LUCI (Labour, Urbanscape and CItizenship) research area. He has been external scientific consultant for UN-ILO International Labour Organization, Project Visiting Professor at the Asia-Europe Institute, State University of Malaya, Kuala Lumpur, Malaysia, and Senior Researcher at Swiss Federal Institute for Vocational Education and Training. His main scientific interest and expertise is in citizenship social and political theories and applied studies on citizenship policies and education, coordinating many European and International research projects in this field.

Contributors

Tuanni Rachel Borba is Researcher in the National Institute of Science and Technology, INCT *Observatório das Metrópoles*. She holds Master's degree in Public Policy, Strategies and Development from the Institute of Economics of the Federal University of Rio de Janeiro (UFRJ). She has experience in research and consulting on projects in the area of gender, transport, and human rights. Her topics of interest also include public policy analysis and theory, urban planning, and urban policy.

Ana Paula Soares Carvalho is Assistant Professor at the Department of Social Sciences of the Pontifical Catholic University of Rio de Janeiro (PUC-Rio) and coordinates the Social Sciences project of PUC-Rio's Institutional Program for Teaching Initiation Scholarships, PIBID. She has experience in the areas of Urban Sociology, Sociology of Law and Political Sociology and Methodology and Practice of Sociology Teaching. She holds a Ph.D. in Sociology from the Institute of Social and Political Studies, IESP-UERJ and Master's degree in Sociology from the University Research Institute of Rio de Janeiro.

Niccolò Cuppini is researcher at the University of Applied Sciences of Southern Switzerland, SUPSI in the LUCI (Labour, Urbanscape and CItizenship) research area. His researches connect urban studies, history of political thought, and critical logistics. He is in the editorial board of Scienza & Politica. He is part of the Into the Black Box research group, he collaborates with the Academy of Global Humanities and Critical Theory, and he is working in many international research projects in Europa, Africa, and Latin America.

Orlando Alves dos Santos Junior holds a Ph.D. in Fundaments of the Urban and Regional Planning by the Federal University of Rio de Janeiro (2000) and Master's in Fundaments of the Urban Planning by the same university. He is currently Director and Full Professor at the Federal University of Rio de Janeiro—Urban and Regional Planning and Research Institute—IPPUR/UFRJ. His research experience is around Sociology, focusing on Urban Sociology, acting in the following issues: democracy, citizenship, urban planning, social participation, and public management.

Larissa Lacerda is researcher and doctoral student in the Sociology Department of the São Paulo University, USP and has a Master's degree in Urban and Regional Planning from the Federal University of Rio de Janeiro (IPPUR/UFRJ) and a degree in Social Sciences by the USP. Her research experience is around the area of urban sociology and urban planning.

Fernanda Amim Sampaio Machado is researcher in the National Institute of Science and Technology, INCT *Observatório das Metrópoles* and a Ph.D. student in the Postgraduate Program at Urban and Regional Planning Institute in the Federal University of Rio de Janeiro IPPUR, UFRJ. She is graduated in Law by the Fluminense Federal University and holds a specialization in Critical Theory of Human Rights by the Pablo de Olavide de Sevilla University (2016) and Master's in Law from the UFRJ. Her experience research focuses on Law and Urban

Sociology, critical theory of law, human rights, right to the city, democracy, culture, sports mega events, urban and regional planning, and critical urbanism.

Humberto Meza is researcher in the National Institute of Science and Technology, INCT *Observatório das Metrópoles* within the Institute of Urban and Regional Planning and Research (IPPUR) of Federal University of Rio de Janeiro. He holds a Ph.D. in Political Science by the University of Campinas (UNICAMP), in the field research of Collective Action, Social Movements, and the interaction with political institutions. He was researcher at the Brazilian Research Center (CEBRAP) in Political Participation and Collective Action.

Patricia Ramos Novaes is researcher in Urban Issues, *Favelas* and the Right to Housing; Gentrification; Neoliberal Urban Restructuring and Public-Private Partnerships. She was Professor in the Public Management Department for Economic and Social Development of the Federal University of Rio de Janeiro, GPDES, UFRJ and is Visiting Professor in the Federal Fluminense University, UFF. She has a Ph.D. and Master's degree in Urban and Regional Planning from the Urban and Regional Planning and Research Institute, IPPUR, UFRJ.

Bruna Ribeiro is researcher in the National Institute of Science and Technology, INCT *Observatório das Metrópoles* by the Urban and Regional Planning Institute of the Federal University of Rio de Janeiro, IPPUR, UFRJ. She has a Master's degree in Urban and Regional Planning and Specialist in Public Policy by the Institute of Economics in the UFRJ. She was consultant analyst at UN-Habitat in the Rio + Social Program—agreement between UN-Habitat and Rio de Janeiro City Hall (2012–2016).

Carolina Pereira dos Santos is researcher in the National Institute of Science and Technology, INCT *Observatório das Metrópoles* by the Urban and Regional Planning Institute of the Federal University of Rio de Janeiro, IPPUR, UFRJ and has a Master's degree in Urban and Regional Planning by the UFRJ and graduated in Economics from the Federal Fluminense University, UFF.

Taiana de Castro Sobrinho is researcher in the National Institute of Science and Technology, INCT *Observatório das Metrópoles* and the Laboratory of Studies on Transformations of Brazilian Urban Law, both linked to the Federal University of Rio de Janeiro, UFRJ. She is graduated in Law from the National Law School of the UFRJ and Master's in Law from the National Law School. Currently, she is member of the Brazilian Institute of Urban Law (IBDU). She has experience in the area of Law, with emphasis on Human Rights, Right to the City, Urban Law, and Criminal Law, with research in right to housing, urban and regional planning, urban policy, human rights, drug policy, and criminology.

Mariana Werneck is researcher in the National Institute of Science and Technology, INCT *Observatório das Metrópoles* and has a Master's degree on Urban Planning by the Urban and Regional Planning and Research Institute of the Federal University of Rio de Janeiro, IPPUR, UFRJ and graduated by International Relations from the Pontifical Catholic University of Rio de Janeiro, PUC-Rio.

Chapter 1
Neoliberalization and Mega Events: The Transition of Rio de Janeiro's Hybrid Urban Order

Luiz Cesar de Queiroz Ribeiro and Orlando Alves dos Santos Junior

Abstract The current developments in Rio de Janeiro in the wake of the 2014 International Federation of Association Football (FIFA) World Cup and the 2016 Summer OlympicGames accelerated the transition from a hybrid urban order to a new urban order of intensified neoliberalization and commodification. This transition is a complex and contradictory process that entails both rollback and rollout features of neoliberalism, as well as a permanent restructuring of sociospatial relations and political arrangements. By discussing the ongoing policies and politics of urban restructuring, we examine the creative destruction of urban structures, institutional arrangements, and regulations of urban space with the transformation of three sites in Rio de Janeiro—Barra da Tijuca, the port district, and the South Zone. The process of variegated neoliberalization generates new modes of production and reproduction, which threaten the principles of democratic governance and universalization of rights to the city.

Keywords Neoliberalization · Urban order · Rio de janeiro

1.1 Introduction

In recent years, the city of Rio de Janeiro has experienced a fast and intense process of urban transformation after hosting a series of sports mega events: the Pan American Games in 2007, the International Federation of Association Football (FIFA) World Cup in 2014, and the Summer Olympic Games in 2016. These mega events have allowed, on one hand, the consolidation of a coalition of political forces at three levels of government (municipal, state, and federal) and major corporations in public works.

L. C. de Queiroz Ribeiro (✉) · O. A. dos Santos Junior
Institute of Urban and Regional Planing, Federal University of Rio de Janeiro, Rio de Janeiro, Brazil
e-mail: lcqribeiro@gmail.com

O. A. dos Santos Junior
e-mail: orlando.santosjr@gmail.com

L. C. de Queiroz Ribeiro and F. Bignami (eds.), *The Legacy of Mega Events*, The Latin American Studies Book Series, https://doi.org/10.1007/978-3-030-55053-0_1

1

On the other hand, they have created favorable conditions for remaking the city with new institutional and spatial arrangements through various neoliberal experiments (Peck and Tickell 2002).

These transformations have been the object of different studies, some of which focus on the city's massive investment in its transportation system (Legroux 2016; Matela 2016), the redevelopment of specific areas such as the port district (Diniz 2014; Werneck 2016), as well as the direct or indirect effects of mega events upon favelas, which have been going through gentrification driven by tourism and land-holding regularization policies (Bonamichi 2016; Cummings 2013). In line with these empirical works, some argue that mega events have intensified and accelerated neoliberalization trends already underway, creating a rupture in the urban order in the institutional, spatial, and cultural dimensions in favor for business interests (Boykoff 2016; Eick 2010).[1]

The interpretation based on the concept of rupture is echoed in works by many others on the impact of the mega events in Rio de Janeiro. For example, Sánchez and Broudehoux (2013) argue that the city's preparation for mega events gave the civic, political, and economic elites legitimacy to transform Rio de Janeiro into a city for capital accumulation. Gaffney (2010) affirms this view by arguing that the mega events have brought about a shift and propitiated the rupture of the legitimizing ideology of the public policy, which increasingly adopts a "shock doctrine" in its discourse, and practice and leaving behind the notion of social inclusion and democratic participation. These interpretations are based on the assumption that neoliberalism is a kind of "economic tsunami," caused by an exogenous force commanded by the global capitalist machinery with an ability to restructure the planetary economic and political geography at various scales (ONG 2007). Though in general we agree with these interpretations, this article advances a different argument.

A deeper understanding of neoliberalization, in our view, should overcome the assumption of an externally imposed economic tsunami. In this article, we conceive neoliberalization as a path-dependent process shaped by particularities of local history and institutions. Different from post-industrial cities in North America and Western Europe, the ongoing restructuring of institutional and spatial arrangements in urban Brazil did not inherit the regulatory landscape or the spatial organization of the Fordist city. Taking from the reference the well-known formulation of Karl Polanyi (2000), the capitalistic spatial patterns of Rio de Janeiro—like in many other Brazilian cities—signal a hybrid logic of urban fabric in which the expansion of the self-regulated market had been taking place without the destruction of the economic dynamic embedded and enmeshed in social relation and social institutions. The city's urban landscape witnesses such hybridity through a complex regime of segregation that combines territorial proximity with social distance between social groups that occupy opposite positions in social stratification.

The mega events-led transformations in Rio de Janeiro need to be interpreted as part of the process of promarket institutional and spatial arrangements, shaped by

---

[1]In this article, we use *neoliberalism* and *neoliberalization* interchangeably, while acknowledging the different treatment of the two terms in the literature.

the context-specific social–spatial formation. Moreover, neoliberalism is not limited to the commodification of the built environment but entails a wider process of transformation of the urban order, such as the creation of a new social milieu favorable to the dissemination and legitimization of promarket values as the basis of social and institutional practices.

This article is based on survey research conducted by the *Observatório das Metrópoles* (Observatory of the Metropolises) network on the impact of the 2014 World Cup and the 2016 Olympics in Brazilian cities. The survey, which took place between 2010 and 2014, monitored budget expenditures and analyzed the social and urban impacts of state interventions in housing, transportation, environment, public security, and sports facilities in all of the cities that hosted these two mega events (Santos Junior et al. 2015a, b). The rest of the article is divided into four sections. In the first section, we briefly review the theoretical literature on neoliberalism, paying particular attention to the shift from the rollback to rollout phase, as well as its contestations. In the second section, we examine how the market and redistributive institutions are intertwined and together shape the uneven development in Brazil. In the third section, we discuss the new cycle of neoliberalization that started in the 1990s and intensified in the 2000s, as exemplified in the commitment made by the city of Rio de Janeiro to host two international mega events. Three examples are used to illustrate this point: the expansion of Barra da Tijuca on the west side of the city as a new strategic node; the recapture of the vast port district in the historic heart of the city by capitalist interests; and entrepreneurial initiatives in the favelas in the southern part of the city, which are manifestations of the rollout phase of neoliberalization. In the conclusion, we discuss the consequences of variegated neoliberalism (Brenner et al. 2010) currently underway in Rio de Janeiro and reflect on the changing forms of urban governance and its possible effects for the future of the city.

1.2 Neoliberalism, Governmentality, and Contestations

The theoretical literature on neoliberalism emphasizes the shift from a rollback phase to a rollout phase in the 1980s, and this transition is manifested differently across localities. After the Thatcher–Reagan era, institutional and spatial promarket arrangements have been put in place through distinct neoliberal experiments. Peck and Tickell (1994, 2002) identified that neoliberal arrangements no longer aim at a rolling back of the state, which involves a phase of intense and accelerated creative destruction of the regulatory framework of the preexisting welfare state, producing immediate effects of the restructuring of sociospatial regulation. This pattern of neoliberalism enhanced the inherent tendencies of a systemic crisis of capitalism and triggered conflicts and contradictions that diminished the social and political legitimacy of the conservative project. The developments in the 1990s are characterized as rollout neoliberalism; that is, "the reconstruction of the forms of the state and its regulatory apparatus, the development of neoliberal governance (moving towards extra market regulation), and a broader 'social' intervention" (Barcellos de

Souza 2013, p. 26). This change of direction has been theorized as a new pattern of the neoliberal project that is disseminated by international organizations through the so-called best practices and fast policies (Peck and Theodore 2015). In addition, this change has made the neoliberal project more dependent on provisional experiments, permeable to contingencies and learning ability of local actors. These perspectives illuminate the concrete processes of neoliberalization by considering historical contexts in which they take place and therefore can help to properly interpret the differentiated effects of neoliberal projects within countries and cities and at different scales.

Studying the institutional and spatial arrangement of the new urban order requires us to conceive neoliberalism as a technique of governing—over populations and territories (Larner 2000). While acting through a constellation of various policies that are not necessarily unified (ONG 2007), neoliberal projects disseminate an entrepreneurial culture and seek to educate individuals to adopt behaviors guided by promarket rationality. In some instances, these projects lead to the destruction of the social, cultural, and territorial foundations that have sustained the institutions of reciprocity and redistribution, whereas in others they propel their reconfiguration. As Peck and Tickell (2002) suggested, neoliberalism has expanded its repertoire of restructuring strategies, by demonstrating a greater capacity to absorb and appropriate local institutions to destroy them in a creative way. In the case of Rio de Janeiro, this has not been different. This can be observed in urban interventions in the central area, as exemplified in the incorporation of popular elements (e.g., Black culture, carnivals, handicraft), simultaneous expulsion (both directly and indirectly) of low-income residents, dissemination of a business culture in the favela, and real estate speculation facilitated by Olympic infrastructural investment.

It is also important to note that, despite the enormously powerful coalition between the government and corporate capital in this recent rollout phase, the ongoing interventions in Rio de Janeiro have been challenged by the civil society, such as professional associations, trade unions, nongovernmental organizations, academic groups, and alternative media, among others. We agree with Peck and Tickell (2002) that the spread of neoliberalism also involves the formation of local networks of resistance, leading to the achievement of social guarantees, even if contrary to its implementation. And as Larner (2000) pointed out, we cannot understand these neoliberalization processes without considering the potential for insurgency by minority groups who constantly challenge the negative legacies of neoliberal restructuring.

1.3 The Hybrid Urban Order of Rio de Janeiro: Different Logics in the Production of Space

The current urban order of the city of Rio de Janeiro has as its main feature uneven patterns of organizational and economic integration, as expressed in the market logic coexisting with principles of reciprocity and redistribution (Polanyi 2000). Classical

Brazilian sociological literature shows that rapid urbanization, associated with indus-trialization and modernization, has not eliminated practices of clientelism, corpo-ratism, or patrimonialism, which still pervade the spheres of the market, society, and state (Diniz 1982; Leeds and Leeds 1980; Nunes 2007; Oliven 2010). Histori-cally, the presence of this form of organization and economic integration produced an institutional pattern of production and reproduction in Brazil, featuring persis-tent informality coexisting with formal modes of production organized by real estate capital and the state. The results of this hybrid institutional pattern are three concomi-tant sociospatial dynamics of use and production of residential space: first, the self-segregation of the upper classes in the form of a strong voluntary concentration in certain well-endowed areas with better infrastructure and urban services; second, the peripheralization of the popular classes in poorer areas with inadequate basic public services (e.g., sanitation, education, health, transport) and far from jobs; and third, the occupation by popular classes of interstitial areas (e.g., hillsides, river banks, idle land owned by public and private entities) outside middle and upper-middle class neighborhoods of the city, a process commonly referred as *favelazation*.

From a morphological point of view, these heterogeneous sociospatial dynamics produce entrenched patterns of residential segregation in Rio de Janeiro, at both micro and macro scales. At the micro scale, territorial proximity is accompanied by social distance between different classes, as evidenced by the presence of favelas in high-income areas. At the macro scale, the continued spreading of informal settlements, developing outside the central city, have widened the gap between working-class neighborhoods and high-income areas. These patterns are shown in Fig. 1.1.

It is important to note that urban informality and illegality—striking features of the sociospatial organization of Rio de Janeiro—are not results of the logic of the needs of the poor not met by the state or by the market (Abramo 2003). It is a more complex process, resulting from industrialization in Brazil in the context of peripheral Fordism and reflecting some of the major modes of governance of the marginal population (Lipietz 1989; Nunes 2007). In this context, commodification of labor occurred through mechanisms such as primitive accumulation of capital (Marx 2013), as well as urbanization of the peasant class. Thus, illegality and land informality are the results of specific forms of urban governance and sociopolitical regulation that sustained Brazilian industrial capitalist development. This is the historical foundation for understanding the urban growth of Rio de Janeiro, which is organized according to similar logics of frontier expansion of capitalist relations.[2] The intensity and speed of mobilization of the labor force via rural-to-urban migration generated major social problems, because the arrival of the rural population was not accompanied by state provision of housing and policy regulation of urban land, as occurred in most Western countries. This process of industrialization through primitive accumulation generated an urban order marked by extensive informal settlements due to precarious public services and infrastructure, as well as irregular land ownership. In this context,

---

[2]*Frontier* here refers to the "boundary that displaces, in its own way, the capitalist mode of produc-tion," expressing "thus, a fundamental contradiction, according to which capitalist accumulation, to move, needs non-capitalist formations around it" (Courlet 1996, pp. 12–13).

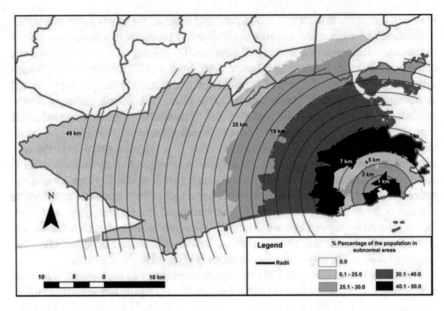

Fig. 1.1 Distribution of the percentage of the population in favelas and precarious settlements by total population, 2010. Data from Instituto Brasileiro de Geografia e Estatística Demographic Census (2012)

traditional logics of urban land commodification coexisted with other logics, such as irregular land settlements, in the territorial organization of Rio de Janeiro.

It is for this reason that in large parts of the territory the dynamic of the urban order built on market mechanisms is weakly regulated by the government and embedded by relations of reciprocity and redistribution through social ties.[3] In Rio de Janeiro, the logics of redistribution and reciprocity are integrated in the market logic of exchange and, as a result, a huge industrial reserve army was uprooted socially in the process of primitive accumulation. In addition, we can infer that these developments led to resistance from the subaltern class, who live in a city where they are unable to fully integrate into the formal labor market and emancipate themselves as working class and are denied basic rights and conditions to exercise their full citizenship. It is from this historical background that we examine the rollout phase of neoliberalism, strongly marked by commodification of urban space and subaltern resistance in the city of striking sociospatial inequality.

The elements presented above show that the current phase of neoliberalization and ongoing commodification in Rio de Janeiro did not originate from the Keynesian or Fordist urban order as in the case of postwar Western Europe and North

---

[3]The literature on migration has demonstrated the role of social networks in the migrants' adaptation process. Singer (1975, p. 55) pointed out the importance of considering this fact by stating that "the adaptation of newly arrived migrants to the social environment often occurs through mutual aid mechanisms and solidarity of older migrants."

America. Therefore, as others have pointed out, neoliberalism should not be understood as monolithic phenomena, imposed unilaterally by exogenous forces to local institutions in each national context considered, but instead should be viewed as a multiscale process resulting from the articulation of broader, global forces with local patterns of economic, political, and social development (Peck and Tickell 2002). Neoliberalization should be treated as both a catalyst and expression of a multiscale process of creative destruction of the existing political and economic space (Theodore et al. 2009). In other words, it should be understood as a global phenomenon that is always variegated in its local manifestations (Peck and Theodore 2015). As Hackworth (2007) stated, "The geography of neoliberalism is much more complicated than the idea of neoliberalism" (p. 11).

Neoliberal urbanization should therefore be understood as a process of permanent destruction and creation of a new economic order of the urban territory, of political institutions directly involved in the management of urban space, and also of the related legal regulations. This process destroys the social relations that support economic exchanges and community lifestyles mentioned above and promote new symbolic representations around the city and the urban. As emphasized by Larner (2000), in order to understand the expansion of neoliberalism, one must consider it not only as a set of changes in the political–institutional framework but also as an ideology shaping collective identities and individual subjectivities, as well as a set of governance strategies that make institutions and individuals conform with market rules. One must take into account "the interactions, dependent on trajectories and contextually specific, that occur between the inherited regulatory frameworks, on the one hand, and projects that emerge from market-oriented neoliberal reforms, on the other" (Theodore et al. 2009, p. 3).

1.4 The 2016 Olympic Games: Inflection Point in the Hybrid Urban Order

In Rio de Janeiro, the process of transition to the neoliberal model of entrepreneurial governance began to manifest itself in the 1990s. From the early 1990s to the late 2000s, the city went through four administrations steered by the ideology of the "competitive city."[4] The city government initiated regulatory reforms and urban restructuring under the guidance of an international network of specialists (Novais 2010). Throughout this period, various experiments solidified the legitimacy of neoliberal policies, providing necessary conditions for the development of the current rollout phase of neoliberalism. Moreover, it was in the political and ideological context of the city's preparation to host the two sporting mega events that the city has entered into a new phase of accelerated neoliberalization. Three factors explain the rise of

---

[4]Mayor Cesar Maia served three terms (1993–1996, 2001–2004, and 2005–2008). Luiz Paulo Conde served as mayor between 1997 and 2000.

neoliberalism in the 1990s and its acceleration in the 2000s. The first stems from positive economic effects of local investment policies under favorable macroeconomic conditions in the country during the period between 2003 and 2014, characterized by real income growth, high levels of employment, and expansion of consumption capacity of the popular classes, as driven by rising international commodity prices and large-scale state investment programs.[5] The second factor corresponds to the consolidation of the governing coalition, by including major national construction companies and the financial sector, as the federal government was determined to launch infrastructural projects for the two mega events in the city. This new coalition of the political, financial, and construction sectors was further strengthened and gained greater legitimacy, which gave rise to a new class of entrepreneurs and business managers. The third factor is the emergence, in society as a whole, and particularly in the grassroots world, of a phenomenon that can be called "business culture," in which economic growth has legitimized promarket solutions to urban problems, discounting the historical debate on the role of the public sector in solving urban problems.

The 2014 FIFA World Cup and the 2016 Summer Olympic Games took place in this context, and they served as strategies to accelerate the neoliberalization process by generating a series of demand to enable institutional reforms and urban interventions. These interventions can be notably observed in the region of Barra da Tijuca, the port district (Zona Portuária), and the South Zone (Zona Sul). To emphasize the economic and social legacies and justify the interventions, the city government called the various projects planned in these areas Olympic projects.

The Olympic project aims largely at a reconfiguration of existing patterns of centrality in the city with interventions in the three aforementioned areas. In the South Zone, the government sought to further strengthen the centrality of this area by directing investment and building infrastructure; in the port area, the government used the Olympics as an opportunity to revitalize the old port and historic downtown; and in Barra da Tijuca, the government wanted to make the area into a new center by massive investment in the real estate sector (Fig. 1.2).

The Olympics-related investment centered on three core policies—transformation, housing, and security. First, for urban transportation, investment was made to build new bus rapid transit (BRT) and light rail vehicle (VLT) routes and also to expand the subway system. Second, in the housing sector, intense real estate investment was poured in the three areas (e.g., Barra da Tijuca, the port, and the South Zone), accompanied by removal of favelas and squatter settlements. Third, public safety was promoted by the state government and centered on the implementation of the pacifying police units (UPP) in selected favelas, in order to reign in organized drug trafficking through militarized control of these neighborhoods. These instances

---

[5]Examples include major infrastructural investment under the Growth Acceleration Program (PAC), investment in social housing promoted by the *Minha Casa Minha Vida* (My House, My Life Program) program, in addition to the significant expansion of activities of the state oil company Petrobras.

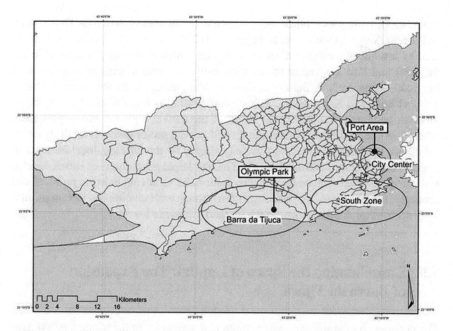

Fig. 1.2 Map of Rio de Janeiro showing three areas of major interventions: the port district, the South Zone, and Barra da Tijuca

of creative destruction of urban structures and socioeconomic dynamics point to a deliberate attempt to reconfigure the city.

These interventions are not random acts but local expressions of a "spatial fix" (Harvey 2013) promoted by neoliberal experiments to create new conditions for capital accumulation in selected areas. This is facilitated by the active role of public authorities, a role that is not limited to make feasible urban renewal projects led by private capital. Rio de Janeiro's city government is the main promoter of the interventions, by articulating and preparing different projects, directly financing various interventions, granting tax incentives and exemptions to attract private enterprises, and adopting new institutional arrangements in addition to changing prior construction sector regulations. The federal and state governments are also deeply involved, particularly with regards to direct investment and financing of urban transportation projects.

Associated with this process of transformation, real estate prices rose quickly, leading to gentrification and uneven development. Here we build on Smith (1987, 2006) and highlight three aspects of gentrification. The first is the class dimension. A shift has been underway in the socioeconomic profile of residents in the targeted intervention areas, from low-income to the middle and upper classes. Second, there is the differential in land rent. Significant price differences exist between current and future land value, especially in those areas that have been devaluated in relation to other areas within the city, thus making these areas attractive for real estate speculation. Examples include areas adjacent to the core of Barra da Tijuca, such as Recreio, Vargem Grande, Jacarepaguá, and Curicica—many of which were occupied by low-income communities and irregular settlements—the port district in the financial and business center of the city; and also some favela neighborhoods in the South Zone, such as Babilônia, Chapéu Mangueira, Pavão Pavãozinho, Cantagalo, Vidigal, and Santa Marta. Third, gentrification is a global strategy for urban renewal, because it is not only the result of the market logic but also represents a strategy of the coalition between public and private sector elites. Overall, gentrification is a complex process and variously shaped by the different local contexts in which it occurs (Janoschka et al. 2014). The municipal government of Rio de Janeiro is directly involved in promoting gentrification, by removing existing political and economic obstacles and by relocating (often through authoritarian and violent means) low-income communities to areas where land prices are lower.

1.5   Consolidating the Space of Capital: The Expansion of Barra da Tijuca

The analysis on Olympics spending shows that a majority of the planned investment (about 62%) is concentrated in the region of Barra da Tijuca (Santos Junior and Lima 2015). These investments—largely associated with public transport projects, expansion of roads, and tunneling—seek to improve the connection of Barra da Tijuca with the South Zone and downtown, thus overcoming the main obstacle for its expansion and transformation into a new business hub. As stated by Guimarães (2015, p. 141), "Massive public investment in accessibility has increased the value of land which for four decades has been concentrated in the hands of the same two owners: Carlos Carvalho and Pasquale Mauro."[6] But the development of Barra would not have been possible without major national construction companies, which were not only responsible for infrastructural works but also actively engaged in real estate development and property management. Prominent among them are Odebrecht Realizações, Andrade Gutierrez, and Queiroz Galvão Desenvolvimento Imobiliário.

As of 2017, the Barra de Tijuca district is predominantly occupied by large gated communities, shopping malls, and corporate buildings. However, several surrounding neighborhoods—such as Jacarepaguá, Curicica, Recreio dos Bandeirantes, and Vargem Grande—are low income and include some favelas. These are the areas that have undergone the greatest change in recent years. The informal settlements and their residents posed huge obstacles to land acquisition, but the Olympics presented a good opportunity because redevelopment was justified by the supposed legacy and

---

[6] As Guimarães (2015) recorded, in 1981 the four major owners of almost the entire landed property of Barra da Tijuca were Pasquale Mauro, Carlos Fernando de Carvalho (owner of the construction company Carvalho Hosken), Tjong Hiong Oei (owner of the company ESTA SA), and Múcio Athayde (owner of Grupo Desenvolvimento). Carlos Carvalho was called by *Veja* magazine the richest man in Brazil ("Os supermilionários," 1981). According to the report, this was the world's largest concentration of urban land (12 million m^2) belonging to a single owner.

enhanced public interest associated with investment in the area. The government opted for evictions, resettling households to peripheral areas through social housing programs subsidized by the federal government, such as *Minha Casa Minha Vida*, and through various compensation arrangements below market prices. The evictions and resettlement can be interpreted as "dispossession" (Harvey 2008), with the transfer of property of the popular classes to other class segments and leading to gentrification in certain areas of Barra da Tijuca.

According to data compiled by the World Cup and the Olympics Popular Committee (Comitê Popular 2015), of the 4,130 families that were removed for Olympic projects, 1,500 families were removed for the construction of new urban transportation systems, especially the Transcarioca, Transoeste, and Transolímpica BRTs. The construction of the Transoeste BRT led to the removal of entire communities, such as Restinga, Vila Harmonia, Recreio II, Notre-Dame, and Vila da Amoedo, totaling approximately 400 families. Thus, in certain areas of Barra da Tijuca, the government helped to promote gentrification by simultaneously removing popular classes and facilitating real estate expansion for higher income groups. As argued by Guimarães (2015),

> The 2016 Olympics will leave as legacy the physical foundations of an economic, political and ideological project that aims to transform Barra da Tijuca into a new centrality in coming decades, deepening the socio- spatial inequalities that characterize the city of Rio de Janeiro. (p. 149)

Despite the mixed nature of urban interventions in Barra da Tijuca, the removal of poor families and the appreciation of real estate value in this region are expressions of variegated neoliberalism in the hybrid urban order of Rio de Janeiro. Barra da Tijuca is still surrounded by many favelas and low-income communities, reproducing, at least in part, the hybrid model of peripheral urbanization that characterizes the city. But significant changes have occurred as the area now features consolidated land parcels and homogeneous neighborhoods inhabited by high-income groups.

1.6 The Recapture of the City Center by Big Business: Redeveloping the Port District

Designated as one of the main legacies, the revitalization of the port district is another neoliberal experiment launched by the city in the preparation for the Olympics, and it represents a recapturing of the city center by capital forces. The Porto Maravilha Urban Consortium Operation was formalized in 2009 and aimed to revitalize the historic port district, an area comprising a total of 5 million m^2 with an overall population of 29,953 in 2010 (Instituto Brasileiro de Geografia e Estatística 2012), whose socioeconomic profile was low- and middle-income. The construction has been carried out by the largest public–private partnership in Brazil, formed in 2010 between the Urban Development Company of the Port District of Rio de Janeiro—which is affiliated with the municipal government—and the winning consortium of the public tender, Porto Novo S/A.[7] Porto Novo S/A will administrate the area for a 15-year period, responsible for providing services and renovation work, in addition to operating and maintaining the area, including the management of public services such as cleaning, lighting, road system, and sanitation.

The revitalization work was financed through the sale of certificates of potential additional construction, which divided the port district into several sectors according to their construction potential, namely the maximum square meters that can be built in each sector. Builders must purchase certificates of potential additional construction, which are also available on the stock exchange for investors. In addition, the revitalization project included heavy investment in historical and cultural heritage preservation—such as the recovery of the Valongo Garden and Wharf, renovation of the José Bonifácio Cultural Center, revitalization of Largo da Prainha and Pedra do Sal, and the construction of large cultural facilities such as the Museum of Tomorrow (*Museu do Amanhã*), the Art Museum of Rio de Janeiro, which were intended to lend symbolic significance to the area.

Given the existence of many favelas, a predominantly low- and middle-income population, and also a major historical and cultural preservation zone, it is unlikely that the port district will undergo a large-scale gentrification process. Thus, the most plausible scenario is strong gentrification in some sections alongside weak gentrification in others. Major real estate and commercial development projects launched in the port district are all located in the areas of highest construction potential, such as the high-income residential project of Porto Vida and the corporate project Torre Carioca Concal.

The newly constructed, and privately managed, residential and corporate buildings coexist and must deal with the presence of favelas, especially favela da Providência, and the tenements, called *cortiços*, which are buildings in precarious conditions with shared bathrooms and kitchens. The tenements were banned by early twentieth century municipal legislations and threatened for demolition during the urban renewal project (1902–1906) led by Pereira Passos, who was inspired by Haussmann's Paris. However, research by the Observatório das Metrópoles shows that there are still a substantial number of tenements in the central area of the city. In the port area alone, 54 tenements are identified, housing more than 1,100 people in rented rooms, which shows the persisting hybrid and peripheral urban order in the city (Santos Junior et al. 2016).

1.7 Building Entrepreneurial Favelas

Since the early 1990s, a process has been underway in Rio de Janeiro that seeks to integrate some of the city's favelas into the market economy by means of a series of

---

[7] The winning consortium of the tender was composed of three construction companies: OAS LTDA, Norberto Odebrecht Brasil S.A., and Carioca Christiani-Nielsen Engenharia S.A.

initiatives taken by the government. Ribeiro and Olinger (2014) stated that a fore-runner of this process was the recognition of these areas by the municipal administration through the *Favela-Bairro* program, initiated in 1988 and financed by the Inter-American Development Bank, which aimed to integrate these areas through specific projects to improve their accessibility and infrastructure. The program was heavily influenced by the populist policies in the 1980s, but in the current rollout phase of neoliberalization, the government has begun to view the favela as a social problem to be solved by discharging these areas from prior social regulations based on principles of reciprocity. Therefore, the government fully incorporated the ideology of the "mystery of capital," formulated by DeSoto (2000) and spread by the international network of think tanks, and launched a land regularization program of favelas (Bonamichi 2016). These transformations are expanded in the context of the preparation for the OlympicGames through favela upgrading programs and promotion of popular entrepreneurship.

In the favelas located in areas characterized by the self-segregation of the upper-middle class, a double trend is observed. First, the representation of some favelas changed from a territory of criminality to places that have been pacified and are now community-friendly, a process that led to the appreciation of nearby property that was previously devalued due to their proximity to the favelas. Second, there has been active promotion of a liberal entrepreneurial culture and of the favela as a commodified space targeting tourist circuits, in the form of bars, restaurants, hostels, and favela tourism. It is a process that has been analyzed as "tourist gentrification," promoted by actors of the formal market in this new frontier for the expansion of capital. Similarly, favelas located in the South Zone have been the object of several upgrading programs. The Babilônia and Chapéu Mangueira hillsides (located in Leme) benefited from the *Morar Carioca Verde* program, which provided various investment such as in public lighting, water and sewage networks, as well as residential buildings. The favelas of Pavão-Pavãozinho and Cantagalo (located between Copacabana and Ipanema), Vidigal (Leblon), and Santa Marta (Botafogo)—all on prime real estate—received *Programa de Aceleração do Crescimento* (PAC) investment and the UPP program. Although these favelas are not the only ones to receive upgrading, there are strong signs that the ongoing interventions in these specific favelas have a gentrifying impact due to their strategic location in the South Zone for big businesses and the Olympic project.

In addition to being the target of land tenure regularization and upgrading, these favelas have received social housing programs, aimed at resettling residents living in risk areas or on streets undergoing expansion, as in the cases of Babilônia, Pavão-Pavãozinho, Cantagalo, and Santa Marta.[8] Such social housing programs, targeting the low-income population, can circumvent the trend of gentrification. In this case, the neoliberal policy experiment is a mixed bag and requires careful analyses.

---

[8]Between 2009 and 2015, 120 social housing units were built in Cantagalo and Pavão-Pavãozinho as part of the Growth Acceleration Program, promoted by the state government of Rio de Janeiro. One hundred seventeen social housing units were built in Babilônia by the City of Rio de Janeiro as part of the *Morar Carioca Verde* program. And in Santa Marta, 64 social housing units are to be built by the state of Rio de Janeiro.

At the center of the ongoing sociospatial transformations, the government and the private sector have also attempted to rebrand favelas as "fashionable space" (Bonamichi 2016) and also ethnic space. Specific initiatives include setting up small business ventures, such as hostels, bars, and nightclubs, to attract tourists and residents from other parts of the city.[9] These favelas paved paths leading to rooftops with spectacular views of the city. One of the examples is the Babilônia favela in Leme, where a path was paved with sponsorship from a major shopping mall in the city (the Riosul Shopping Center).

These various initiatives were aimed at turning the favela into a symbolic asset to raise property value in areas of already high real estate prices (e.g., Botafogo, Leme, Copacabana, Ipanema, and Leblon). A case in point is the well-known Vidigal favela, subject to an accelerated process of social transformation driven by the nongovernmental organizations and medium and large real estate companies operating in the wealthiest areas of the city.[10] Moreover, Vidigal is one of the few favelas in the South Zone with no social housing programs. With a spectacular view to the sea, this favela, or at least part of it, might effectively be undergoing gentrification. If we compare real estate prices of three selected areas—Leblon, one of the most expensive neighborhoods in the city; Vidigal, close to Leblon; and Leme, which houses in its interior the Babilônia favela—all three areas appear to have undergone strong real estate valuation after the installation of UPP programs in Vidigal and Babylon in 2009.[11] In addition, a significant rise in property prices is observed in the Vidigal favela and also in the Leme neighborhood, perhaps reflecting the impact of upgrading programs and branding strategies for Babilônia. This is not observed in the case of Leblon, perhaps because the Vidigal favela is not located in the heart of this neighborhood, thus having a lesser impact on the valuation of properties in Leblon. The drop in real estate prices in 2014 in all analyzed locations can be explained as a result of the economic crisis in Brazil from 2014 onward (Table 1.1).

Beyond the dynamics of real estate valuation, one must consider another aspect associated with the ongoing transformations; that is, promoting businesses within favelas. The neoliberal experiment appears to be strongly associated with the commodification of the actual experience of visiting the favela, thus spreading the values of liberal entrepreneurship into this space. In this case, this represents not so much the change in the social composition of these territories but the spread

---

[9]Favela Babilônia was the setting of a Brazilian soap opera aired by the country's largest television network, Rede Globo, in 2015.

[10]It is exemplary for three reasons. First, on the one hand, there is a typical process of accumulation by dispossession based on appropriation and reinterpretation of the collective cultural capital as a mark of the social distinction of the commodified spaces by the logic of the Income Art (Harvey 2005). Secondly, for expressing the action of the transfer and adaptation mechanisms of the neoliberalization policy mentioned in the literature (De Soto 2000). Finally, it is an experiment in constructing the social consent necessary to consolidate the progress of neoliberalization expressed by practices of symbolic violence on the former collective imaginary of the city, strongly established in the 1980s around favelas as spaces to be integrated via the extension of citizenship.

[11]Information on real estate prices in the Babilônia favela is unavailable.

Table 1.1 Real estate valuation per square meter (M^2) of Leblon, Leme, and Vidigal Favela, 2008–2015

Month of reference	Leblon		Vidigal		Leme	
	M^2 value (R$)	Valuation compared to 2008 (%)	M^2 value (R$)	Valuation compared to 2008 (%)	M^2 value (R$)	Valuation compared to 2008 (%)
September 2008	11,252.49	xx	2,372.74	xx	7,069.09	xx
September 2009	12,408.30	10.27	3,161.63	33.25	7,618.79	7.78
September 2010	14,982.28	33.15	7,050.49	197.15	10,780.19	52.50
September 2011	16,310.29	44.95	6,317.33	166.25	12,464.37	76.32
September 2012	15,997.13	42.17	8,870.68	273.86	12,923.54	82.82
September 2013	15,322.74	36.17	9,318.15	292.72	13,490.36	90.84
September 2014	13,973.17	24.18	9,991.72	321.10	12,928.38	82.89
September 2015	13,386.00	18.96	8,757.00	269.07	11,657.00	64.90

Note Data on Vidigal and Leblon for September 2011 are based on numbers for October 2011, due to the absence of information on Vidigal for September of that year
Source FipeZAP (2015); the figures are updated and tabulated by the authors, based on ORTN/OTN/BTN/TR/IPC-r/INPC

of a liberal promarket culture that replaces former social structures based on reciprocity. The dissemination of this entrepreneurial culture has been happening through microbusinesses training and funding programs—promoted by public agencies, nongovernmental organizations, and private companies—and also through the spread of enterprises catering to favela tourism, such as hostels, bars, restaurants, and tour operators. Entrepreneurial discourses and practices introduced a new form of self-government to the local population (Foucault 2004), gradually replacing clientelistic networks and also repression that characterized the relationship between the state and these territories in previous eras.

There have been many examples of fostering community entrepreneurship. The first was the creation of the Community Entrepreneur Award by the State of Rio de Janeiro Promotion Agency[12] in 2003, which was given to 13 businesses financed by its Microcredit Program in the pacified favelas. A resident of the Vidigal favela—an owner of a hostel—was the winner in the "Successful Business" category. Other award categories also reflected the dissemination of entrepreneurial, promarket culture, such as "Sustainable Business," "Young Entrepreneur," "Entrepreneur of the Visual Communication Industry," "Enterprising Woman," and "Innovative Business." It is worth highlighting that the facilitators involved in promoting these businesses in the favelas also received awards (Agência Estadual de Fomento 2013). The second example was a survey on entrepreneurship in favelas carried out in 2015 by

---

[12]The State of Rio de Janeiro Promotion Agency is a semipublic corporation created by Federal Decree No. 32376 on December 12, 2002 (Governo do Estado do Rio de Janeiro 2002). It is affiliated with the Rio de Janeiro State Department of Economic Development, Energy, Industry and Services with the aim of stimulating economic development.

iMaster (2015) and publicized by the second New Brazilian Favela Forum—a partnership between the Data Popular Institute and Unified Favela Union (CUFA).[13] At the forum, the business interest among favela residents was especially emphasized. According to the survey, "There are 3.8 million people in Brazilian favelas willing to start businesses" and "in general, the entrepreneurial population is female, black and young." The survey also notes that "among those who want to open their own business, 63% want to do it within the favela itself" (iMaster 2015).

Finally, there was the example of Favela Holding (FHolding), which, according to its website (www.fholding.com.br), "is a group of companies whose main objective is the development of favelas and their inhabitants. FHolding is active with community entrepreneurs, fostering and promoting new business opportunities, entrepreneurship and employability." Its goal by 2020 "is to be recognized as the world's largest business promoter in favelas." Among the companies linked to Favela Holding include *Favela Vai Voando*, founded in partnership with the company Empresa Vai Voando to open sale offices of airline tickets, with affordable prices to residents; Avante, a financial solution agency that aims to "humanize financial services," empowering favela residents to solve their financial problems; Data Favela, the first research and business strategy agency that studies consumer behaviors in favela communities and identifies business opportunities for companies wishing to expand their activities in favelas; Favela Shopping, which promotes new ventures in favelas focused on generating income and opportunities; and League of Community Entrepreneurs, a national cooperative of entrepreneurs involving groups of favela businesses, street vendor communities, and the fishing community, among other groups.

Unlike the urbanization initiatives practiced in the favelas during the 1980s and 1990s—resulting from the intermingling between the logic of the market and the logic of reciprocity and representative of an imaginary of a social integration based on the expansion of citizenship and urban infrastructure—recent housing policies seek to promote a process of disembedding these territories from local institutions, with the promotion of tourism and commercial activities and by spreading urban entrepreneurship logics among its residents.

These new practices replace the old discourse, focused on citizenship rights, with a new one focused on entrepreneurship and educating the poor to treat their (now valued) spaces as business and thus turning the favela into a commodity. In addition, these new programs spread the idea of individual responsibility on which business success or failure is dependent. The new ideology of business in favelas is one of the most powerful forces of the current rollout phase of neoliberalization underway in Rio de Janeiro.

---

[13]CUFA is a nongovernmental organization created over 20 years ago, by youth from favelas in Rio de Janeiro. CUFA promotes various activities in favelas, such as education, leisure, sports, culture and citizenship, literature, and other social projects. One of its founding members is the rapper MV Bill, who has received several awards for his participation in the hip hop movement. In 2015, CUFA opened an office in New York (Fundação Doimo 2015).

1.8 Conclusion

This article has interpreted the current urban transformation in Rio de Janeiro as a process of variegated neoliberalization, strongly driven by the organization of the sports mega events, especially the 2016 Summer Olympic Games. The transformation marks a rollout phase of neoliberalization and involves creative destruction of urban spaces, institutional arrangements, social regulations, as well as massive financial and policy interventions in certain areas of the city, particularly in the regions of Barra da Tijuca, the port district, and the South Zone. To conclude, we highlight some of the major consequences of variegated neoliberalization on urban governance and the future of Rio de Janeiro.

First, it is important to recognize the scales at which this process of variegated neoliberalization unfolds. Brazil is experiencing a paradoxical situation in the period herein considered. On one hand, at the national level, since 2003—first with the Lula da Silva administration and later with the Dilma Rousseff administration—the federal government has adopted a neodevelopmental economic policy in which the state heavily intervenes in the economy with the implementation of new Keynesian policies[14] to rebuild the conditions of capital and labor circulation in more socially fair terms. These policies stimulate demand through the creation and expansion of consumer credit, income transfer, and minimum wage increase. On the other hand, at the local level, neoliberal projects proliferated as strategies for competitive integration of cities in the globe economy. This paradox marks the specificity of variegated neoliberalism in Brazil and has numerous consequences for development.

Second, it is worth noting that the changes associated with variegated neoliberalization both preserve old agents, practices, urban structures, and institutions inherited from earlier eras and incorporate new actors, practices, and institutions. There is neither a complete rupture that introduces an entirely new pattern of urban development nor the mere continuity of previous processes. The current variegated neoliberalization embodies inherited political structures and is characterized with path dependence. The situation presents challenges to understand how the old and new political structures are intertwined and how they co-produce an emerging hybrid urban order. From the point of view of urban governance, neoliberal policies reinforce conventional practices of urban accumulation in the history of Rio de Janeiro and shun away from democratic management associated with ideas of the "right to the city" (Harvey 2012; Lefebvre 1967). In this context, public participation is progressively replaced by decision-making processes that submit the government to the market logic.

---

[14]Highlights among these Keynesian policies include the *Bolsa Família* program, which ensures a minimum income for low-income families; the PAC, aimed to bolster urban infrastructure; and the *Minha Casa Minha Vida* program, focused on new housing construction.

Third, the implementation of the neoliberal project entails contradictions and leads to conflicts, opposition, and resistance. Examples include issues regarding investment priorities, forced evictions, and the lack of channels for social participation. Such conflicts—involving a variety of organizations and social movements—can influence the course of the neoliberal project, changing it substantially or even rendering it impracticable, depending on how strong they will become over time. These developments introduce uncertainty about the future of the city. Urban governance in Rio de Janeiro will be marked by the intensification of conflicts.

Finally, we cannot ignore the strength of the power coalition behind the neoliberal project. It has demonstrated great ability to incorporate many secondary interests and to create alliances with other agents and political structures in order to facilitate the implementation of the neoliberal project. Overall, variegated neoliberalization involving both old and new political structures is threatening the principles of democratic governance and the universalization of rights to the city. This calls for an opposition block to the ongoing transformations, in order to overcome the fragmentation produced by the power coalition and to build an inclusive, fairer, and more democratic city.

References

Abramo P (2003) A teoria econômica da favela: Quatro notas sobre a localização residencial dos pobres e o mercado imobiliário informal [The economic theory of the favela: four notes on the residential location of the poor and the informal real estate market]. In: Abramo P (ed) A cidade da informalidade: O desafio das cidades latino-americanas [The city of informality: The challenge of Latin American cities]. Sette Letras/Fundação de Amparo à Pesquisa do Estado do Rio de Janeiro, Rio de Janeiro, Brazil, pp 189–224

Agência Estadual de Fomento (2013) Prêmio Empreendedor da Comunidade: Vidigal leva a melhor [Community entrepreneur award: vidigal leads the best]. https://www.agerio.com.br/index.php/portal-pld/161-01-11-14

Barcellos de Souza M (2013) Variedades de capitalismo e reescalonamento espacial do Estado Brasileiro [Varieties of capitalism and space resettlement of the Brazilian State] (Unpublished doctoral thesis). State University of Campinas, São Paulo, Brazil

Bonamichi NC (2016). Favela on sale. Regularização fundiária e gentrificação de favelas no Rio de Janeiro [Favela on sale. Land regularization and gentrification of favelas in Rio de Janeiro] (Unpublished master's thesis). Federal University of Rio de Janeiro, Rio de Janeiro, Brazil

Boykoff J (2016) Power games: a political history of the Olympics. Verso, New York, NY

Brenner N, Peck J, Theodore N (2010) Variegated neoliberalization: geographies, modalities, pathways. Global Netw 10:182–222

Comitê Popular da Copa e Olimpíadas Do Rio de Janeiro (2015). Megaeventos e violações dos direitos humanos no Rio de Janeiro—Dossiê do Comitê Popular da Copa e Olimpíadas do Rio de Janeiro. Olimpíada Rio 2016, os jogos da exclusão 2015 [Mega-events and human rights violations in Rio de Janeiro—Dossier of the popular committee of the world cup and olympics in Rio de Janeiro. Rio 2016 Olympics, the 2015 exclusion games]. https://issuu.com/mantelli/docs/dossiecomiterio2015_issuu_01

Courlet C (1996) Globalização e fronteira [Globalization and frontier]. Ensaios FEE, Porto Alegre 17:11–22

Cummings J (2013) Confronting the favela chic: gentrification of informal settlements in Rio de Janeiro, Brazil (Unpublished master's thesis). Harvard University, Boston, MA

De Soto H (2000) The mystery of capital: why capitalism triumphs in the West and fails everywhere else. Basic Books, New York, NY

Diniz E (1982) Voto e máquina política: Patronagem e clientelismo no Rio de Janeiro [Voting and political machine: patronage and cronyism in Rio de Janeiro]. Paz e Terra, Rio de Janeiro, Brazil

Diniz N (2014) Porto Maravilha: Antecedentes e perspectivas da revitalização da região portuária do Rio de Janeiro [Porto Maravilha: Background and prospects for the revitalization of the port region of Rio de Janeiro]. Letra Capital, Rio de Janeiro, Brazil

Eick V (2010) A neoliberal sports event? FIFA from the Estadio Nacional to the fan mile. City 14:278–297

FipeZAP (2015). Fundação Instituto de Pesquisas Econômicas - Fipe and ZAP Portal [Foundation Institute for Economic Research – Fipe and ZAP Portal]. https://www.zap.com.br/imoveis/fipe-zap-b/

Foucault M (2004) Sécurité, territoire, population: Cours au collège de France, 1977–1978 [Security, territory, population: courses at the Collège de France, 1977–1978]. Gallimard/Seuil, Paris, France

Fundação Doimo (2015) CUFA lança escritório em Nova Iorque (EUA), e cria Semana Global [CUFA launches New York office, creates global week]. https://www.fundacaodoimo.org/site/2015/09/14/cufa-lanca-escritorio-em-nova-iorque-eua-e-cria-semana-global/

Gaffney C (2010) Mega-events and socio-spacial dynamics in Rio de Janeiro 1919–2016. J Latin Am Geogr 9:7–29

Governo do Estado do Rio de Janeiro (2002) Decreto Nº 32.376 de 12 de Dezembro de 2002 [Decree No. 32,376 of December 12, 2002]. https://www.scribd.com/document/293173807/Decreto-32376

Guimarães RCV (2015) Barra da Tijuca e o projeto olímpico: A cidade do capital [Barra da Tijuca and the Olympic project: The city of capital] (Unpublished master's thesis). Institute of Urban and Regional Planning at the Federal University of Rio de Janeiro, Rio de Janeiro, Brazil

Hackworth J (2007) The neoliberal city: governance, ideology, and development in American urbanism. Cornell University Press, New York, NY

Harvey D (2005) A produção capitalista do espaço [The capitalist production of space]. Annablume, São Paulo, Brazil

Harvey D (2008) O neoliberalismo: história e implicações [Neoliberalism: history and implications]. Loyola, São Paulo, Brazil

Harvey D (2012) Rebel cities. Verso, London, England

Harvey D (2013) Os limites do capital [The limits of capital]. Boitempo, São Paulo, Brazil

iMaster (2015) Pesquisa revela dados sobre empreendedorismo nas favelas [Research reveals data on entrepreneurship in favelas]. Portal iMaster, p.1. https://imasters.com.br/noticia/pesquisa-revela-dados-sobre-empreendedorismo-nas-favelas/

Instituto Brasileiro de Geografia e Estatística (2012) Censo Demográfico 2010 [Demographic Census 2010]. Ministério do Planejamento, Orçamento e Gestão, Instituto Brasileiro de Geografia e Estatística

Janoschka M, Sequera J, Salinas L (2014). Gentrification in Spain and Latin America—A critical dialogue. Int J Urban Reg Res 38:1–32

Larner W (2000) Neo-liberalism: policy, ideology, governmentability. Stud Polit Econ 63:5–25

Leeds A, Leeds E (1980) A sociologia do Brasil urbano [The sociology of urban Brazil]. Zahar, Rio de Janeiro, Brazil

Lefebvre H (1967) The right to the city. In: Kofman E, Lebas E (eds) Writings on cities. Blackwell, London, England, pp 63–184

Legroux J (2016) O Rio de Janeiro como sede da Copa do Mundo 2014 e das Olimpíadas 2016. Marketing metropolitano, projetos de transporte e impactos sobre as iniquidades socioespaciais [Rio de Janeiro as host of the 2014 World Cup and the 2016 Olympics. Metropolitan marketing, transportation projects and impacts on socio- spatial inequalities] (Unpublished doctoral thesis). Université de Lyon 2, Lyon, France

Lipietz A (1989) Fordismo, fordismo periférico e metropolização [Fordism, peripheral Fordism and metropoliza- tion]. Ensaios FEE 10:303–334

Marx K (2013) O capital: Crítica da economia política: Livro I. O processo de produção [Capital: Critique of political economy: Book I. The production process]. São Paulo, Brazil: Boitempo

Matela IP (2016) A gestão dos transportes: A renovação do pacto rodoviarista [Transport management: the renewal of the road pact]. In: Ribeiro LCQ (ed) Rio de Janeiro: Transformações na ordem urbana [Rio de Janeiro: Transformations in urban order]. Letra Capital, Rio de Janeiro, Brazil, pp 515–543. https://transformacoes.observatoriodasmetropoles.net/livro/rio-de-janeiro/

Novais P (2010) Uma estratégia chamada "planejamento estratégico" [A strategy called "strategic planning"]. Editora 7 Letras, Rio de Janeiro, Brazil

Nunes E (2007) A gramática política do Brasil [The Brazilian political grammar]. Jorge Zahar Editora, Rio de Janeiro, Brazil

Oliven RG (2010) Urbanização e mudança social no Brasil [Urbanization and social change in Brazil]. Centro Edelstein de Pesquisas Sociais, Rio de Janeiro, Brazil

Ong A (2007) Neoliberalism as mobile technology. *Transactions of Institute of British Geographers*, *32*, 3–8. Os supermilionários [The supermen]. (1981, June 17). Veja 667:52–53

Peck J, Theodore N (2015) Fast policy experimental statecraft at the thresholds of neoliberalism. University of Minnesota Press, Minneapolis

Peck J, Tickell A (1994). Searching for a new institutional fix: the after-Fordist crisis and the global–local disorder In: Amin A (ed) Post-fordism: a reader. Blackwell, Oxford, England, pp 280–316

Peck J, Tickell A (2002) Neoliberalizing space. Antipode 34:380–404

Polanyi K (2000) A grande transformação: As origens da nossa época [The great transformation: The origins of our age]. Campus, Rio de Janeiro, Brazil

Ribeiro LCQ, Olinger M (2014). The favela in the city-commodity: Deconstruction of the social question. In Ribeiro LCQ (ed), The metropolis of Rio de Janeiro. A space in transition. Observatório das Metrópoles/Letra Capital, Rio de Janeiro, Brazil, pp 19–36

Sánchez F, Broudehoux AM (2013) Mega-events and urban regeneration in Rio de Janeiro: planning in a state of emergency. Int J Urban Sustain Dev 5:132–153

Santos Junior OA, Gaffney C, Ribeiro LCQ (eds) (2015) Brasil: Os impactos da Copa do Mundo 2014 e das Olimpíadas 2016 [Brazil: The impacts of the 2014 world cup and the 2016 olympics]. E- papers, Rio de Janeiro, Brazil

Santos Junior OA, Lacerda L, Werneck M, Ribeiro B (2016) *Projeto Prata Preta: levantamento de cortiços da área portuária do Rio de Janeiro* [Silver Black Project: survey of tenements of the port area of Rio de Janeiro]. Rio de Janeiro, Brazil: Observatório das Metrópoles. https://www.observatoriodasmetropoles.net/images/abook_file/pratapreta2016.pdf

Santos Junior OA, Lima CGR (2015) Impactos econômicos dos megaeventos no Brasil: Investimento público, participação privada e difusão do empreendedorismo urbano neoliberal [Economic impacts of mega-events in Brazil: Public investment, private participation and diffusion of neoliberal urban entrepreneurship]. In Santos Junior OA, Gaffney C, Ribeiro LCQ (eds) Brasil: Os impactos da Copa do Mundo 2014 e das Olimpíadas 2016 [Brazil: The impacts of the 2014 World Cup and the 2016 Olympics]. Rio de Janeiro, Brazil: E-papers, pp 57–78

Singer P (1975) Economia política da urbanização [Political economy of urbanization]. Brasileira, São Paulo, Brazil

Smith N (1987) Gentrification and the rent gap. Ann Assoc Am Geogr 77:462–465

Smith N (2006) A gentrificação generalizada: De uma anomalia local à "regeneração" urbana como estratégia urbana global [Generalized gentrification: From a local anomaly to urban "regeneration" as a global urban strategy]. In: Bidou-Zachariasen C (ed) De volta à cidade. Dos processos de gentrificação às políticas de "revitalização" dos centros urbanos [Back to town. From the processes of gentrification to the policies of "revitalization" of urban centers]. Annablume, São Paulo, Brazil, pp 59–87

Theodore N, Peck J, Brenner N (2009) Urbanismo neoliberal: La ciudad y el imperio de los mercado [Neoliberal urbanism: The city and the market empire]. Temas Sociales, Santiago De Chile 66:1–11

Werneck MGS (2016) Porto Maravilha: Agentes, coalizões de poder e neoliberalização no Rio de Janeiro [Porto Maravilha: agents, coalitions of power and neoliberalization in Rio de Janeiro] (Unpublished master's thesis). Institute of Urban and Regional Planning at the Federal University of Rio de Janeiro, Rio de Janeiro, Brazil

Chapter 2
When the Lights Turn off. Rio de Janeiro's Wannabe Global City Trajectory

Niccolò Cuppini

Abstract When the famous Saskia Sassen's book "The Global City" was published (1991), she mentioned only three cities as effectively global cities (New York, London, and Tokyo). However, in the last few decades, we have assisted to a proliferation of global cities worldwide, as points of the emerging network of global interconnections. This phenomenon implies a complex assemblage of institutions and different actors. Becoming a global city means elaborating a texture of convergence between State and Regional institutions, private capital, and global actors (multinational corporations, global institutions, financial institutions), as well as an active role "from below" (migration to the city, civil society limelight, local entrepreneurship etc.). Therefore, at the swing between the Millennium, when the so-called BRICS counties started to emerge, each of them invested in creating their own global cities. This article investigates the strategy adopted by the Brazilian State and its articulation to transform Rio de Janeiro into a Global city, particularly focusing on the decade of global events hosted within the city from 2007 to the Olympics of 2016. Moreover, the analysis focuses on the first legacy of this strategy of urban transformation.

Keywords Global city · Mega events · Urban conflicts

2.1 Introduction

Cities are increasingly becoming crucial nodes of the shifting geographies of the contemporary world. They represent strategic spaces for economic and cultural innovation, financial management, and governance strategies, as well as for the possibility of gaining "presence" for excluded and marginalized people. Cities are the concrete

N. Cuppini (✉)
Department of Economics, Health and Social Sciences, DEASS - Labour, Urbanscape and CItizenship, LUCI Research area, University of Applied Sciences and Arts of Southern Switzerland, SUPSI, Lugano, Switzerland
e-mail: niccolo.cuppini@supsi.ch

L. C. de Queiroz Ribeiro and F. Bignami (eds.), *The Legacy of Mega Events*, The Latin American Studies Book Series, https://doi.org/10.1007/978-3-030-55053-0_2

places that drive globalization processes, concentrating people, commodities, capitals, and languages within themselves while at the same time connecting global flows. Since the 1980s, the deep interconnection between cities on the global scale started to represent a new trans-national space, the so-called global city (Sassen 1991). Being part of this new global space is quintessential for every territory aspiring to participate in contemporary economy and political dynamics.

When Saskia Sassen's famous book "The Global City" was published (1991), she mentioned only three cities as effectively global cities (New York, London, and Tokyo). However, in recent decades, we have witnessed a proliferation of global cities worldwide, as points of an emerging network of global interconnections. This phenomenon implies a complex assemblage of institutions and different actors. Becoming a global city means elaborating a texture of convergence between state and regional institutions, private capital, and global actors (multinational corporations, global institutions, financial institutions), as well as an active role "from below" (migration to the city, civil society limelight, local entrepreneurship etc.).

Therefore, at the turn of the millennium, when the so-called BRICS (Brazil, Russia, India, China, South Africa) countries started to become more influential globally, each of them invested in creating their own global cities. This chapter investigates the strategy adopted, at both the national and local level, within the Brazilian case to transform one of its biggest cities, Rio de Janeiro, into a global city, focusing on the coalitions of actors that prompted and improved this perspective, their urban logic, and the legacy of this strategy. The most important vector of this process has been the massive concentration of "mega events" in the city, particularly in the decade from 2007 (the Pan American Games) to the Olympics of 2016. Most of the literature tends to focus on the project and implementation phases of those events, while obliterating their long-term 'results' (Baade and Matheson 2015; Gaffney 2015). Rio de Janeiro, as well as many other cities, has been widely studied during these mega events in the making, especially with respect to the hosting of the World Cup in 2014 and the Olympics in 2016 (Müller and Gaffney 2018; Maturana-dos-Santos and Pena 2017). However, it is rare to find in-depth studies of these cities when the lights of these events turn off.

Rio de Janeiro's global image, forged through a typical city-branding strategy, has often been presented as a renewed sparkling metropolis. However, the research on the ground has shown a series of serious, profound contradictions, and the huge crisis of that model of development. The political scenario is quickly mutating and polarizing, as the Brazilian general elections of October 28, 2018, which turned the political arena upside down, has shown. If we look at the situation in Rio de Janeiro, some historical dynamics are coming back on the scene: urban violence has re-erupted with thousands of killings per year, and emblematically on September 2, 2018, the major pieces of Brazil's scientific and cultural heritage went up in smoke, as a devastating fire ripped through much of Rio de Janeiro's Museu Nacional. Therefore, "we must say that twenty-first century Rio de Janeiro remains the same. A city-metropolis that reproduces in its core the unequal urban model set when it merged into the second wave of the national peripheral modernization started in the 1930s", wrote de Queiroz Ribeiro (2017b: 2). *Plus ça change, plus c'est la même chose*, it could be added. The injection of huge amounts of capital within the city, driven by logics and dynamics of global actors, has reproduced the limits and contradictions of an urban development "with steroids" (Davis 2006), with its bubbles of development and its hard social and political problems when the bubble explodes.

To elaborate on these topics, this chapter presents a historical background of Rio de Janeiro development, and then describes the project of transforming it into a global city through the analysis of its globalizing strategies and the implementation of mega events. At the end of the chapter, a perspectival analysis of the legacy of these processes will be discussed, focusing mainly on the question of the transportation system and public security. This research is based on a literature review of Rio de Janeiro development and more broadly on mega events and globalizing strategies (mostly in the Global South), on the setting of a data set and a quantitative analysis of the urban transformations of Rio de Janeiro's last decade, and two months of fieldwork research with 24 in-depth semi-structured interviews targeting relevant actors of Rio de Janeiro public administration and institutions, private actors, and social movements, in order to understand who the subjects and powers at play are, in the setting up of its urban regime (Lauria 1996; Kantor et al.1997).

2.1.1 A Global History

Rio de Janeiro's history has been "global" since its origins. Founded in 1565 by the Portuguese, Rio remained as the capital of the pluri-continental Lusitanian monarchy from 1808 until the War of Brazilian Independence (1822). It served as capital of the Empire of Brazil until 1889, and then the capital of the Brazilian Republic until 1960, when the capital was transferred to Brasília. Rio was a city of colonizers and enslaved people from Africa, being the largest port of slaves in America. This tumultuous history was consubstantial with a lack of a major urban planning, until the first huge urban intervention promoted by Pereira Passos, named mayor in 1902. His explicit project was to build a new Paris on the Atlantic, taking inspiration from the profound works of transformation that radically reshaped the French capital in the second half of the nineteenth century. The "modernization" of the city implied the demolishing of the *cortiços*, where most of the poor population lived (mostly descendants of slaves). This population then moved to live in the city's hills, creating the first wave of favelas, which are still a landmark of Rio. Moreover, Passos brought electric power to Rio, adapted the city to automobile circulation, and built many important buildings. These urban reforms promoted by the ruling classes to build the tropical Paris are another example of the global roots of Rio, but this global imaginary accompany all the city's history (think of the postmodernist architecture of the works in Barra de Tijuca—to create "the Miami of Latin America"—or the most recent turn in global technological urban development imaginary (Di Bella and Rossi 2017).

Moreover, Rio's history has also been heavily influenced by political dynamics, above all the moving of the nation's capital city to Brasília—one of the most important global modernist projects, an ex novo city realized in 1960. When in 1964 a *coup d'état* installed a military dictatorship, the city-state of Rio was the only state left in Brazil to oppose the military. However, in 1975 a presidential decree removed the city's federative status, merging it with the State of Rio de Janeiro, and Rio de Janeiro replaced Niterói as the state's capital. It was the birth of the Rio de Janeiro Metropolitan Region (RJMR), which has been heavily impacted by the loss of its industrial supremacy over the last few decades (moving from Rio to Sao Paulo) (Teixeira 2015).

Between the 1980s and 2010s, the RJMR went through a growing urbanization process disconnected from the consolidation of an economic base to sustain it. At the same time, as shown by Ribeiro (2017b), this "metropolisation" has been characterized by three key urbanizing movements that affected the interaction of "center-periphery", as happens in all the Brazilian metropolitan regions: a kind of "auto-segregation" of economic elites inside highly valued spaces (i.e. South Zone in Rio City and Barra da Tijuca), the expansion of the "center to periphery" (middle-income people that migrated to peripheral municipalities such as Niteroi, Sao Gonçalo or Nova Iguaçu) and the "peripherization" of the center, associated with the "favelization" of the metropolitan core areas and/or the precariousness of popular sectors (Jardim 2007). Those dynamics featured the trajectory of the RJMR over four decades, experiencing several urban configurations. The current RJMR is the second most important metropolitan region in Brazil in terms of demographic weight and political influence and the third in Latin American countries. According to CEPERJ data,[1] the RJMR is integrated by almost 12.2 million people and a GDP of approximately R\$ 405 billion (USD 126.6 billion). This represents almost 64% of all the State GPD. Although the RJMR configuration suggests a kind of centrality (Rio City) and 20 peripheral municipalities, this metropolitan region realizes a typical "sponge-effect" (Nuvolati 2011): the centralities absorb population during the day and expel them during the night.

During the period of Brazilian democratization in the 1980s, Rio suffered for the depletion of the industrial model, which contributed to cause a vertiginous growth of poverty, inequalities, and violence. This dynamic strengthened the historical trend of richer people leaving the city center, a phenomenon that somehow anticipated the North American "white flight" of the post-War. Having some urban centralities as de-valorized areas is a crucial step to project a global city, because these areas are low-price zones inhabited by marginalized populations, and therefore ideal bases to develop new urban speculative and financial projects (Abu-Lughod 1999). The two successive phases of the national political economic arena (the neoliberal policies of the 1990s and the neo-developmental policies of the following decade) followed a global and continental trend, posing two other decisive elements to pursue the project of making of Rio a global city.

---

[1]CEPERJ is public Office for statistical and economical research that depends of the State of Rio de Janeiro. For more information, you can see: https://www.ceperj.rj.gov.br/.

2.2   Wannabe Global City

Brazil was at the forefront of the so-called BRICS (Brazil, Russia, India, China, South Africa) countries in the first years of the new millennium. The groundwork for this positive economic conjuncture was laid during the late 1990s, with the project to transform Rio de Janeiro into a global city. When the "wannabe global city" began to emerge, many interviews describe the typical dynamic at the start of every global city-formation: the wealthiest classes buy into the historical city center at a very low price, expelling its population and transforming it into a new urban scenario. However, compared to other processes of global city formation, the specificity of Rio seems to be the more complex social landscape and the resilience of its lower classes. The image of the "dual city" (Mollenkopf and Castells 1992) is not completely applicable to Rio. There are many overlapping layers to consider in order to understand Rio's socio-urban shape, a multiplicity of dualizing processes in many directions. We have the favelas and the skyscrapers facing each other, a Rio peculiarity within the global scenario, but at the same time the favelas of the Southern Zone are "class A favelas", compared to those of the Northern Zone. How to "deal with" favelas is a long-term question for every city administration, but when the transformation of the "center" of Rio got underway according to new global trends, a new logic and rationality was adopted to apply the global city project. First of all, there was a "logistical rationality" at play, another common element of every global city building. As Mauro Osorio, President of the Pereira Passos Institute (the Urban Planning Institute of the Rio Municipality), explains:

> "The logic of the removals of the 60s is very different from the logic of the removals of Eduardo Paes [Rio's Major from 2009 to 2016]. In the 60s, the poor were taken from the South Zone to value the South Zone, the favela left the chick to enter a middle-class housing complex. This was not the case in Eduardo Paes' s management. In his administration, virtually all removals were to make new transportation circuits".

It is a circular logic, implying that the urban texture has to be made flexible and adaptable, a characteristic directly derived from the "financial logic" endowing the formation of global cities (Cuppini 2015). There is another resonance with the typical historical discourse to transform the urban tissue worldwide, recalling especially the work of Hassmann in nineteenth century Paris, that of "sanitation":

> "all this starts from the beginning of the creation of UPPs [Unities of Pacifiying Police], which at least for me, were born to deterritorialize those people who were in the communities. This has to do with the sanitation processes of Rio de Janeiro, which happened with Mayor Pereira Passos, at the beginning of the twentieth century, and in reality, at the time of the real estate explosion, this hygiene happened in Providencia, it happened in Vidigal, in Dona Marta and Cantagalo, of real estate speculation within the comunities themselves".
>
> (Roberto Gomes, coordinator of the Central de Movimentos Populares (CMP))

From the end of the 1990s the idea of making Rio a global city began to spread. One important factor was Brazil's increased capacity to control inflation. Before that there was a deep crisis. 1994 was an important year: the Brazilian Real, the present currency, was created, leading to an important stability and to "the possibility, for the first time, to imagine and plan the future. In 1995–1996 Brazil had a huge development, a change in mentality. It was possible to dream and the will to show Rio to the world emerged, and to start planning for new development. Secondly the was the question of oil, allowing for a lot of money to be made. Rio is closely linked to oil, thanks to Petrobras, which is based in Rio and pays its taxes here. There was a lot of money, and people thought anything was possible, a collective euphoria. If you had a project, Rio was the place to be".

(Luis Gabriel Denadai, Vice Director of Urban Planning within the Municipal Department of Urban Planning and Housing (SMH) of the Municipality of Rio de Janeiro)

However, the most important strategy for globalizing Rio has undoubtedly been the massive investment in hosting mega events, one of the most typical urban development strategies of the recent decades (Hiller 2004). Particularly from 2007 to 2016 we have witnessed an impressive wave of events taking place in the city, particularly from 2007 to 2016, in an attempt to profoundly reshape its urban and social texture. However, all these changes had to deal with the historical configuration of the city:

"The Pan American games are the moment when new cases of evictions started. Before that they were episodical and mostly about real estate speculation. Now they have become systematic, a project of the city thanks to this global plan for the city. The removals are different from the past. Before the discourse was: "This occupation is illegal", and the government made the eviction. Now the discourse is about environmental questions, the Olympics etc.".

(Jailson Silva, founder of the Observatorio das Favelas)

This seriously changed the morphology of the city, even considering that half of the buildings in the city of Rio de Janeiro are illegal, either because they are in favelas, or because even if they are not in favelas, they are not regularized, being either self-constructions or irregular constructions on regular lots. In sum, profound widespread social/urban change began in the 2000s, with the administrative strategy and the problems enlightened by the words of Denadai and Antonio Carlos Barbosa, president of CDURP, the Urban Development Company of the Port Region:

"In this region that we call Potential Centrality, we stimulated the densification, the uses of commerce and services, new niches, and we returned to the zoning of Rio de Janeiro. We must remember that technically it is from 1976, but in fact it is based on an understanding of the city that comes from the previous legislation of the 1960s, that is to say, it is very modernist with that segregated vision of the city which mandates that in this corner is the dwelling, in that corner the industry, in that corner the workplace, etc., but with a desire to expand the city, to gain space. Therefore it is done with the expansion of Rio in mind, which is when the city began to grow with the arrival of migrants, who experienced an expansion of barriers, etc."

"The decade of global events has shown how outdated Rio's legislation is, its weakness. We have to consider that this period was when the city started to concretely come out of the society of the Seventies that still existed, when Brazil really started to change, passing from a rural to an urban society. People used to think: "Now we are a country of urban development. We are urban, industrial, and modern. At least we made o developed society, now it is time to think about post-industrial development". There started the idea of the mega events, but within an old legislative framework. This problem was underestimated because a lot of wealth was circulating".

What the qualitative research has shown (see also Ribeiro and Rojas 2014) is that the elaboration and application of a globalizing strategy in Rio has been built upon a logic structured on a peculiar financial and logistical pattern, but we will come back to that. What is remarkable is the evidence that this strategy, even though these main characteristics are somehow "new", is marked by the recurrence of historical trends, continuing to influence current urban development, as in the example of the rhetoric on "sanitation" and the continuous re-emergence of the "question of the favela" have demonstrated. This may explain the persistence of the multiple dualizing processes that still characterize Rio. Moreover, various interviews point to the fact that the juridical and administrative frameworks of the metropolis were not "prepared" to sustain such strategies. As a result, when the boost of the mega events ended, the "institutional machine" had neither instruments nor competences for a rapid re-orientation of its agency.

2.3   Globalizing Rio

Even though we have shown how the history of Rio de Janeiro has always been "global", in recent decades there have been quantitative and qualitative differences from its global connections that should be taken into account. Above all, globalizing the city has become, in recent decades, a precise strategy of neoliberal economic growth at both the urban and the national level (Marcuse and Van Kempen 2011). In fact, the global city concept is not an urban concept: the global city is not a city as such, but rather a global network of production. It illustrates how the activities of production are scattered on a global basis, in the era of globalization. These globalized production networks require new forms of financial and production services to manage them, subject to agglomeration economics, tending to cluster in a limited number of cities. Global cities are

> "particular 'post-industrial production sites' where innovations in corporate services and finance have been integral to the recent restructuring of the world-economy now widely known as globalization. Services, both directly for consumers and for firms producing other goods for consumers, are common to all cities of course; what we are dealing with here are generally referred to as advanced producer services or corporate services. The key point is that many of these services are by no means so ubiquitous".
>
> (Beaverstock et al. 1999: 126).

Therefore, even within a large variety of methodologies, approaches, and classifications, what has emerged is a complex panorama of many global cities distributed worldwide and with specific internal and external hierarchies and degrees of political and economic dependence. Consequently, to compete at the global level every territory needs to develop each proper global city, as Brazil did. If a global city should be an advanced producer services node, an economic giant, an international gateway, and a political and cultural hub, the Brazilian system invested in differentiated ways in its metropolis to achieve such results. It is remarkable to note that, if we follow the ranking developed by the periodical bulletin of the Globalization and World Cities (GaWC) Research Network—https://www.lboro.ac.uk/gawc/gaw cworlds.html, where cities are assessed in terms of their advanced producer services using the interlocking network model (see Taylor 2010) to measure a city's integration into the world city network, we observe this dynamic: São Paulo has constantly remained at the top level from 2000 to 2018 (Alpha); in 2000 Belo Horizonte and Porto Alegre were scored with "Sufficiency" but in 2018 passed, respectively to Gamma—and High Sufficiency; Brasilia entered for the first time in the ranking in 2018 (Sufficiency); Rio de Janeiro in 2000 was scored as Gamma+, in 2010 passed to Beta, and in 2018 is Beta. Therefore, within an overall improvement of all Brazilian global cities in this ranking, Rio is the one that made the greatest improvement. The quantitative and qualitative research done within the project "Urban Regimes and Citizenship" in Rio de Janeiro somehow confirms this trend, although also shedding light on many crucial factors that remain "off the map" (Robinson 2008), when we analyze cities only from the above-mentioned perspective. Thus, we will try now to show the logic of urban development, its actors, the variables that have been taken into consideration, and finally the legacy of this process for a wider social spectrum.

2.3.1 Logic of Urban Development

What has distinguished Rio from the other Brazilian metropolises is, first of all, the number of mega events that Rio hosted. While the impressive flow of capital invested to prepare the city for these events has been key in its transformation, the logic behind these transformations must also be fully grasped. In fact, all the narratives sustaining the huge public investment in the mega events and the correlating urban transformations are based on the reference to other global cities' iconic transformations ("Looking at the other countries' experiences, Barcelona created a company to play a project of this size, London created a company for that, Buenos Aires also in Porto Madero, which is what inspired us to approve the CDURP"—Alberto Gomes, former president of CDURP between 2011 and 2016). But more generally, what is the leading logic that has oriented the coalition of Rio's urban regime? From many interviews it is possible to deduce, as a hypothesis, that there is a typical financial rationality at play, leading the public/private governance toward "expectations" and an "enduring growth" as the main vectors around which to shape urban development.

Alberto Gomes indirectly discussed this topic, frequently referring to urban transformations in terms of market transformations ("The real estate market has another rhythm"; "CEPAC prices are at the market price; they vary depending on the availability of purchase. Well, the Fund went into business, the wheel started turning. It doesn't sell, it negotiates. It's like, "I have a banana and you a stall. I'll give you the bananas and when you sell, we'll split the profit" That's what they're doing… So the problem, in my view, is not the model, but the ability of these guys (FGTS-FIIPM) to play the business"). Annie Amici, Manager of the Urban Mobility Department of the National Economic and Social Development Bank (BNDES) mobilized the question of "confidence" ("What happens in the Rio de Janeiro State? The State of Rio had its accounts unbalanced, and because of this even if it had the authorization to go into debt, there would probably be no one to offer it resources because if its accounts are unbalanced and has no ability to pay, it does not generate confidence"), Luis Gabriel Denadai said. "At Porto Maravilha, the idea was for the market to do everything. The idea was to take this zone, transforming the urban space of that zone by selling constructive potential".

Alvaro dos Santos Pereira (Professor within the research "Public Account-ability for Residents in Contractual Urban Redevelopments—PARCOUR") affirmed "It was an urban operation that was based on credibility"; the Eunice Horacio, Urban Mobility Manager by the FETRANSPOR (Federation of Passenger Transport Companies of the State of Rio) sustained that "all the rationale sustained the urban transformations were based on expected performances of added value".

Basically, all the interviews presented Rio's urban evolution as based on expectations, trust, confidence, and faith in the market, stretched on a temporality of gaining "profits" in the future: a typical financial logic that often works, thanks to the monetary support by the public, who sustains these procedures, as the Porto Maravilha case illustrates (see the chapters on it within this book). Therefore, an urban development is possible only thanks to "additives":

> "It means that you enter in the competition with a basic project and win the bid to run this basic project, perform the work. That is, the competitor (the company) does the project, does the work and does everything. And what did the guys do? They got additives. You win the competition with the price you offered, then you start the work, but inform them that you would face a difficulty of execution with the costs presented, and have to have an addition. These are the additives".
>
> (Deputy in the Rio Legislative (ALERJ) Eliomar Coelho Neto of PSOL (Socialism and Liberty Party)

We could consider the concept elaborated by Mike Davis of an urban development "with steroids", where the strategy of the mega events has worked precisely in this direction, shaping a "bubble" that has been able to "modernize" the city in terms of investments in infrastructures, but at the same time now Rio is living the "bursting" of the bubble with dramatic social consequences, as we will see. In any case, all the previously mentioned conditions (economic growth, the stability of the Real, the constitution of a coalition leading the urban regime, etc.) would probably not have been sufficient, per se, to sustain the global city's formation without the resources that only mega events can attract, from national and international capital. That is the basic idea underlying the tremendous effort to collect the series of events from 2007 to the Olympics in 2016.

However, what emerges from the interviews carried out and from part of the literature, it would be untrue to affirm that those who invest in the construction of these events are "big capital". The role of the public is crucial in creating a reliable environment for investments, in paying the preliminary tranches for the startup of the events, in generating socio-economic interest in the promotion of the city as an actor capable of winning the hosting of the events, in defining the partnerships with the private sector, in designing the urban planning according to the projects for the events, and so forth. Therefore, who pays for the bet on the mega events? It is the public who invests in such a bet: "But whoever paid for it and continues to pay is the taxpayer. Everything falls on the back of the worker" (Deputy Eliomar Coelho Neto). As stated before, the question is not only about finance and economy, but there is also a huge political theme at stake. Effectively, "the adoption of the "global city" category requires us to observe the institutional convergences (of local, state and federal interfaces), as well as the interactions of the various agents (state, private capital, multinational corporations, social movements) in the maintenance of the current urban order" (Hoyler et al. 2018: 181).

2.3.1.1 Actors

"There was an articulation built on top of a tripod: the public power in its three instances (federal, state and municipal), the big contractors representing the great economic interests, and the media that went inside their house every day and every hour".

Deputy Eliomar Coelho Neto

In setting up an urban regime, the stabilization of definite coalitions is a core theme (Mossberger and Stoker 2001). What emerges from the analysis is that it is very difficult to configure such conditions of a long-term cooperation pattern due to the continuous variations within and among the different variables. This leads us to suppose that in the last decade the Rio de Janeiro metropolitan area has developed only through temporary coalitions, partially coordinated and partially conflicting. However, there is a crucial scale question involved in this reflection. If we focus on the coalitions on the scale of the Rio city or of the Rio state, the image emerging from the interviews is that of an articulated block of power, but with crucial capabilities to coordinate and elaborate complex strategies, a coalition built up during the 1990s with a specific developmental strategy based on mega events, rapid economic growth based on extractivism (see Mezzadra and Neilson 2017) and the attraction of big capital at national and international levels. What emerges is a political, economic, and medial pact able to align the federal, state, and municipal governments apart from their political "color", composing around them some economic sectors, especially the touristic and construction sectors, sustained by an explicit coverage by the O' Globo network, the main media power of Rio, and somehow supported by the judicial power. This network of powers made coalitions during the 1990s and the early 2000s, strengthened by the winning of the mega events, by a marketing/branding strategy, and by a positive economic conjuncture, leading to a consolidated coalition during the first decade of the new millennium, also thanks to a strong intervention to reduce crime rates.

The coalition was formed at the national level, but Rio became the laboratory of this pact and the Brazilian international window of it. The "result" is a "very crazy decision process" (Osorio), with many "grey areas", that have, however, been able to orient the city's development, figuring out a pyramidal schema of the urban regime as sketched below. This configuration broke down during 2013–2016; due to the 2013 protest movements, the biggest in Brazilian history, recognized by many interviewees as a turning point (see also Cocco and Moulier Boutang 2013; Vicino and Fahlberg 2017), the economic crisis led by the fall of the prices of raw materials such as oil and the consequent crisis of the extractivist model; the international conjunction, the crisis of BRICS; and the political-juridical crisis, with the eruption of many corruption cases. Therefore, from the end of the Olympics to now, there has been a significant weakening of the power coalition that led Rio for almost 20 years, from 1994 to 2014. Many people say the situation has returned to that of 20 years ago, with a resurgence of violence and instability. Another thing should be added: most interviewees stressed the decisive role of the political apparatus, which seems to be the leading force of the coalition.

It is possible to trace a panorama of the actors of Rio's urban regime, starting from our interviews and the literature analysis. They can be divided into four categories: 1. legitimate public actors; 2. legitimate private actors; 3. semi-legitimate public and private actors; 4. other actors. Here the term "legitimate" is adopted to intend the actors that are formally (institutionally/legally) involved in urban policies, as distinct from other actors who concretely intervene, but without a formal frame for their actions.

1. Legitimate public actors: Three governmental levels—national, state, municipal; public security (mainly the UPP, political parties, and city ministries);
2. Legitimate private actors: the private stakeholders involved in the PPPs, Public Private Partnerships. However, "it is difficult to clearly understand the economic side of the companies; their role is often not explicit and many things move in the shadows". What should also be added is the role of big financial capital, national and international, which is, however, quite elusive and difficult to precisely identify;
3. Semi-legitimate public and private actors: public and private actors not directly connected to democratic procedures linked to the intervention on the urban transformations, but in fact influencing them. The judicial power and O' Globo seem to be the most important. A more in-depth look should be made of the role of the BNDES, a bank that seems to have a "political" role in making decisions about the financing of the projects; BNDES is a federal and public bank;
4. Other actors: there are many minor actors within this field that are not directly connected to the decision-making process of the city, but who influence it in many different ways: contesting, supporting, influencing the urban regime coalition. They can be informal or "illegal" actors: the already mentioned social movement of 2013; the *trafico* groups, mainly Terceiro Comando and Comando Vermelho; the *milicia* (see De Paula 2015); the lobbies, the public fetranspor. "Rio de Janeiro has a very late political logic, as it has little regional reflection and lobbies always have a very big weight. And I'm talking about the FETRANSPOR lobbies, the incorporating sector that ends up having more importance in Rio de Janeiro";

many agencies acting as "technical" actors supporting other actors, often inter-laced with public security. It would be possible to add other "collective actors" like tourists and "it's a lot of money, they can even buy ministers. It's a Mafia" (Osorio).

Given this general overview, it is necessary to add that the "real government" of Rio (in terms of what Michel Foucault would have named as "governamentality") seems to be a govern of the exception, a management of the interstices between right and anomy, licit and illicit, state and illegal powers (Cava). These thresholds cross the entire city, defining its citizenship regimes. Therefore, what emerges is a dynamic cartography of interpenetrations, shades, strategies and grey zones, rather than a map of unified actors.

From the interviews it is possible to hypothesize that the Olympics in particular are a perspectival point around which a series of processes have been aggregated. The Olympics were an accelerator and the final step of the decade of global events, whose continuous insertion of "additives" should have been transformed into a definitive and stable transformation of Rio into a global city. But the legacy is more confused and the end of the story still has to be written:

"On the one hand, Porto Maravilha is placed as a legacy of the games for the city. At the same time, the realization of the games is mobilized as a backdrop to generate optimism and boost the realization of the works in the port. The games, in a way, will bring investments to the city, and it's obvious that such a project can absorb all the investment demands. Paes used to say that the games have become a justification for everything. So the Olympics offer a context favorable to the argument of urgency. Use this speech to make special legislation for the project, do everything quickly, linking the three spheres of government with flexibilization of contracting mechanisms and removals. All this had to be ready for the Olympics, taking away the social barriers that accompany this project" (Zacone).

This quote is quite emblematic of the hypothesis about the Olympics as an accel-erator, but there is something more here. Considering the fact that in the beginning of 2018 the military was put in charge of Rio public security, what is the relationship of this event with the Olympics, if any? A hypothesis that we could launch and work on could be that the Olympics was used as an "experiment" of a "state of exception" (O'Hara 2017; Schausteck de Almeida and Bastos 2016) or "state of emergency" (Sanchez and Broudehoux 2013), as a tool to *force* the "democratic procedures". Therefore, from the Olympics to the military's new role in 2018 one could trace a continuity rather than a discontinuity.

2.4   The Legacy of the Globalizing Strategy

"Cities like Rio de Janeiro, Johannesburg, Mexico City, and Santiago have registered some of the highest levels of inequality in the world. Emerging Gateway cities need to address these challenges if they wish to continue their growth trajectory"

(Trujillo and Parrilla 2016: 31)

In recent years, the state of Rio de Janeiro (and particularly the city of Rio, as its most important urban center) has experienced traumatic institutional moments. Among them is the declaration of a "state of calamity" at the beginning of 2017 with a financial deficit of almost R$ 18 billion; the accentuation of violence in the municipalities of the metropolitan region, particularly police violence, responsible last year for 17% of the deaths in the state; and a growing weakening of the metropolitan institutionality, despite the Metropolis Statute in force since 2015. Keeping in mind our objective to trace the trajectory of Rio's wannabe global city, it is crucial to keep the research open on what happened in the aftermath of the decade of global events, considered as the crucial strategy for globalizing Rio. Too often there is a good amount of research on the construction and on the events themselves, but when the lights go out even the eyes of the research community tend to follow the flows of global events rather than to elaborate independent research prospects and more accurately, in-depth analyses of the complex impact that they produce. Apart from the daily news reporting on the serious problems emerging in Rio in recent years, most of the interviews we realized affirmed that the legacy of the globalizing strategy was almost a failure in terms of collective gains for the majority of the population. The phrase "we have gone back to ten years ago" recurred frequently.

"History is repeating itself: every day we have a shooting somewhere. Every day there is a child without a school. Every day there are people forbidden to cross a street because they hear a gun shot. This is not common in other metropolises", affirmed Azis Filho (communication entrepreneur), and this "ghost" of urban violence is resurging with some "spectral" results in the attempts of urban transformation: "As a result of the crisis of 2014, 2015 and 2016, residential companies stopped building in Rio. Nowadays almost nothing is built in Rio. And the commerce itself stopped [….]. The skeleton of the building is still there" (Alexandre Mendes, researcher). Urban administrators and technicians sustain that it is now necessary to radically re-frame the ways of intervention: "Money is lacking for everything now. Either we try to play with the natural dynamics of society or we won't be able live in the same place because we don't have the money the requalification of a region, something like what was done in the Port, where public power paid for everything and then people arrived. We don't even dream of making such a change today" (Denadai). The absence of capital for urban transformations after the euphoria of the global events is not only related to the economic crisis that Brazil is experiencing or to the loss of global capital investments (that have dried up without leaving long-term benefits), but must also be connected to the fact that the public has had a crucial role, not only via political decisions, but also in financial terms. As the specific analysis of the Porto Maravilha regeneration project (which was almost totally state-driven) shows, it is possible to assume that a political break is one of the most important causes of the disruption of this kind of development, along with the economic crisis. It is also valuable to add that many interviews reported Porto Maravilha, as well as Barra de Tijuca, to be models of urban development adopted from other experiences (mainly Barcelona, Boston and Baltimore.)

However, apart from the handbook knowledge, it is relevant to note that only the "positive effects" have been considered by the stakeholders, as the president of CDURP emblematically stated: "Are you aware of the social costs and negative effects that similar projects of urban regeneration have had worldwide?"; "Mmm... Yes I know more or less, but I am sure this is a different story". Finally, from this story it is remarkable to note that the vision of the city has always started from specific points rather than from a holistic perspective.

This raises the point of the scale of analysis and intervention, and to conclude this chapter there are two main topics that are necessary to tackle as the most relevant legacy of Rio's "wannabe global city" and to understand the future perspectives for the development on the metropolitan scale: security and transportation.

The resurgence of "war" within the city after the Olympics has been a decisive term of the public debate, a sort of perverse legacy of the decade of mega events. Before the military intervention in Rio in 2018, many interviews realized during 2017 stated that: "There you have this very shallow perception that public security is a police matter, and when the police officer doesn't solve it because the whole system is weak, it calls in the Army" (Zacone, policeman), "the central problem is security" (Robson, Rodrigues is a colonel of the Military Police of the State of Rio de Janeiro (PMERJ), "I think as long as it does not solve this situation of insecurity, no project will work in this region" (Carlos Hermanny Filho, Director of Engineering, Sustainability and Innovation, Odebrecht), "I think the problem is security; we can invent anything, but as long as there is a security problem, nothing is going to work" (Annie Amici, Manager of the Urban Mobility Department, BNDES). The relationships between public security, the global events and their legacy, considering that a "strong" intervention of the state has been one "necessary" aspect to make the hosting of the events possible, but also the fact that, as two interviews sustain, there was a sort of "pact" between legal and illegal actors to pacify the city during the events: the more the business works, the more there is an interest in doing business rather than war.

Furthermore, many interviewees say that the transportation system is the best positive legacy for Rio after the decade of global events (Kassens-Noor et al. 2016). However, there are also many critical voices about that, based on the classical question: "For whom?"

"Today there's a feeling that the city flows better than it did five years ago. There is a feeling, right? The big question that we have regarding the issue of mobility is that first the public power doesn't play its part and delegates all of this to the business community. The public power should take some of this in their hands. [...] The entire road network built during the mega events, all of it, was all delivered to private initiative. You have nothing of public power. Nothing", affirms Reimont Otoni, councilor of the city of Rio de Janeiro, the Workers Party (PT), 2016–2020. It is not by chance, therefore, that "the project of the Olympic games in Rio have privileged the wealthiest central zone, the Zona Sur and Barra de Tijuca, with really few interventions in the metropolitan area" (Gomes), which is the poorest one. There is a logistical power (Cuppini 2017) at play, elusive but really crucial at the same time: "The transportion companies are in fact governing Rio, and they are contrary to investing in something other than buses. Transportation is not sufficient, and there is a political block that does not allow for a quality transportation system. There was a project to construct a subway for the peripheral areas, but there is a political decision in the public/private relationship that is stopping it, that is not developing a thought/project for the whole city" (Alvaro dos Santos Pereira, researcher), "There is a way of governing the city through the management of the transportation system, that is entirely a political question" (Otoni). This "political power" of transportation is deeply intertwined with the production of the "multiplicity of dual cities" characterizing Rio's social and urban landscape, and it is something that will be needed to be seriously taken into account for future investigations focused not only on global flows, but also by examining how flows of people can or cannot move within urban spaces.

2.4.1 What Scale?

A more general question can be posed. When we analyze how global dynamics and projects impact a city, defining the very "boundaries" of the city itself is somehow a political choice. In this sense, if we analyze Rio the city or the Rio metropolitan area we would articulate very different conclusions in evaluating the effects of the decade of global events on Rio. As we have seen, if we focus on the first scale, many indicators suggest a positive effect, while very little has been said and studied on the second scale. The analysis shows the profound institutional fragmentation of the metropolitan area of Rio de Janeiro and the absence of regulatory institutions of public services in the metropolis. What emerges is a highly uneven development of the metropolitan area, leading us to hypothesize that the configuration of urban regimes in the city of Rio de Janeiro has historically neglected the metropolitan scale (for a general reflection see Brandão et al. 2018).

However, it is quite evident that there is not a unique pathway of development in the metropolitan area, as this territorial scale has never really been considered in local or national policies (in terms of metropolitan governance and urban management). The legacy of the global events on the metropolitan area is very weak, irrelevant or even pejorative in some cases. The infrastructuralization of the metropolitan area remains scarce, and "urban planning and the informal expansion of the city go in different directions" (Denadai). The Rio metropolitan area has evolved mainly due to the agency of *os pobres* and informal and illegal actors, within a landscape that is increasingly *fraturada* (as many interviewees reported). The expansion of the metropolitan area, moreover, is also the result of the impact of the global events: the re-location of many people removed from their houses to make space for the infrastructures extended the city, as Gizela Martins, communications adviser for the Rio de Janeiro Assembly (ALERJ) in the Human Rights Commission, explains:

"In 2009 news broke out in the city's major newspapers that 119 favelas would suffer with the removals. And these residents learned about these removals from the newspaper itself. [...] Obviously, this news in 2009 came out days after Rio became the host of the mega-events. Then Rio became the host of the World Cup, Olympics, everything. And a week later the newspapers announced these removals. That is, everything was planned. [...] Many of the people removed were in the West Side of Rio. They were in Minha Casa, Minha Vida houses, which is problematic because they are places where they don't have a hospital. Where there is no bus. There is no sanitation. There is no school. They are abandoned in these places and in addition many of these places are militia sites. […] And these villagers are also losing their homes due the militia actions. That is to say, it destroys the conviviality of a favela. […] And 11,000 families who were removed, evicted as well, were promised social rent. They have been receiving it for a while, but today they don't have this social rent anymore because of the crisis. […] And there was the news of the rich Rio de Janeiro,the Rio de Janeiro that was all in peace, a Rio de Janeiro that would advance technologically, that would improve the life of everyone. And there we have today, Rio de Janeiro, after the mega events, ten years of mega-events finished, schools closed, teachers without receiving salaries, public universities closed".

Therefore, the question of the legacy of the global events in the metropolitan area is quite evident, as Vicente Loureiro, the Executive Director of the Metropolitan Chamber for the Governmental Integration of the Rio de Janeiro State, explains: "So, that is to say, a very fractured region, both from the social and urbanistic point of view, as of the installed institutional capacity to solve problems. […] The imbalance is so great", he continues, giving the example of Japeri, a town 70 kms from the core zones, very poor and with the worst HDI in the region, a dormitory city employed in small local commerce, subsistence, and proximity services "without life" and with very little industrial employment. 90% or more of the economically active population has to move to work, almost always to Rio. 50%, or nearly so, of the daily travel needs in the metropolitan territory are for services, access to health and education, and the other half is for work. It is therefore clear that "having 85% of the hospital beds concentrated in the city of Rio de Janeiro is unsustainable […] So I think from the metropolitan point of view the Olympics, in the case of the metro, was harmful to the metropolitan region. I don't think it was good". Loureiro then continues asking: "Who decides these kinds of priorities? The state of Rio. It was the state together with Rio City who made feasible the attractiveness of the Olympic Games. The process of seduction, right? For the conquest of the Olympics, to show a connectivity project for that concentration there, I think it reversed the priority and from my point of view for a construction of a more equal metropolis and with better opportunity distributed in the territory, was not good. It was not the best choice".

To conclude, this chapter shows how the globalizing strategy applied to Rio by the elite coalitions has reproduced, and somehow improved, the historical inequalities characterizing the city. The mega events empowered the economic performance of the city in terms of global connections or within the richest zones, but displaced (Ren 2017), commodified (Ribeiro and Junior 2017) and dispossessed many parts of the population (Bin 2017; Vannuchi and Van Criekingen 2015). This is due to political decisions, but it is also entangled in Rio's history and in the persisting global inequalities that tend to make the global city strategy more fruitful for global players than for local populations. The post-Olympic scenario is dark: the state of Rio declared it was broke and was rescued by a government bailout in 2016; Rio state's former governor Sérgio Cabral is in jail for corruption, like many other politicians and companies involved in the Olympics (the Petrobras scandal and Lava

Jato operation); urban violence is sharply increasing along with unemployment and impoverishment. In the end, the Olympics were a success in television and in Barra de Tijuca (an upscale west Rio suburb of condominiums, malls and freeways), but working-class *Cariocas* and lower-income areas did not benefit, or eventually were damaged by it.

References

Abu-Lughod JL (1999) New York, Chicago, Los Angeles: America's global cities. University of Minnesota Press, Cambridge

Baade R, Matheson V (2015) An analysis of drivers of mega-events in emerging economies. Holy Cross Working Paper Series

Beaverstock J, Smith RG, Taylor PJ (1999) A roster of world cities. Cities 16(6):445–458

Bin D (2017) Rio de Janeiro's olympic dispossessions. J Urban Aff 39(7):924–938

Brandão CA, Ramiro Fernández V, Ribeiro LCQ (eds) (2018) Escalas Espaciais, Reescalonamentos e Estatalidades: Lições e desafios para América Latina. Letra Capital, Rio de Janeiro

Cocco G, Moulier Boutang Y (2013) The first revolt of metropolitan labor multitude. Multitudes 54(3):19–31

Cuppini N (2015) Towards a political theory of the globalized city. Scienza & Politica XXVII(53):247–262

Cuppini N (2017) Dissolving Bologna: tensions between citizenship and the logistics city. Citizsh Stud. https://doi.org/10.1080/13621025.2017.1307608

Davis M (2006) Fear and money in Dubay. New Left Review 41:47–68

De Paula LA (2015) The "Grey Zones" of democracy in Brazil: the "Militia" phenomenon and contemporary security issues in Rio de Janeiro. Spatial justice, UMR LAVUE 7218

Di Bella A, Rossi U (2017) Start-up urbanism: new York, Rio de Janeiro and the global urbanization of technology-based economies. Environ Plan A 49(5):999–1018

Gaffney C (2015) Gentrifications in pre-olympic Rio de Janeiro. Urban Geogr. https://doi.org/10.1080/02723638.2015.1096115

Hiller HH (2004) Mega-events, urban boosterism and growth strategies: an analysis of the objectives and legitimations of the Cape Town 2004 Olympic Bid. Int J Urban Reg Res 24(2):449–458

Hoyler M, Parnreiter C, Watson A (2018) Global city makers. Economic actors and practices in the world city network. Elgar, New York

Jardim AP (2007) Pensando o espaço e o território na metrópole do Rio de Janeiro. KG & B, Ribeirão Preto

Kantor P, Savitch HV, Hadock SV (1997) The political economy of urban regimes: a comparative perspective. Urban Aff Rev 32:348–377

Kassens-Noor E, Gaffney C, Messina J, Phillips E (2016) Olympic transport legacies: Rio de Janeiro's bus rapid transit system. J Plan Educ Res: 1–12

Lauria M (1996) Introduction: reconstructing urban regime theory. In: Lauria M (ed) Reconstructing urban regime theory: regulating urban politics in a global economy.SAGE Publications

Marcuse P, Van Kempen R (eds) (2011) Globalizing cities: a new spatial order? John Wiley & Sons, New Jersey

Maturana-dos-Santos LJ, Pena BG (eds) (2017) Mega events footprints: past, present, and future. Engenho, Rio de Janeiro

Mezzadra S, Neilson B (2017) On the multiple frontiers of extraction: excavating contemporary capitalism. Cult Stud 3(2–3):185–204

Mollenkopf JH, Castells M(ed) (1992) Dual city, restructuring New York. Russel, New York

Mossberger K, Stoker G (2001) The evolution of urban regime theory: the challenge of conceptualization. Urban Aff Rev 36(6):810–835

Müller M, Gaffney C (2018) Comparing the urban impacts of the FIFA world cup and olympic games from 2010 to 2016. J Sport Soc Issues 42(4):247–269

Nuvolati G (ed) (2011) Lezioni di sociologia urbana. il Mulino, Bologna

O'Hara J (2017). https://stateofexception.com

Sanchez F, Broudehoux AM (2013) Mega-events and urban regeneration in Rio de Janeiro: planning in a state of emergency. J Int J Urban Sustain Dev 5(2):132–153

Ren X (2017) Special issue: urban transformations and spectacles in Brazil. J Urban Aff 39(7):893

Riberio LCQ (ed) (2017a) Urban transformations in Rio de Janeiro. Development, segregation, and governance. Springer, Cham

Riberio LCQ (ed) (2017b) Rio de Janeiro: Transformações na ordem urbana. Letra Capital, Rio de Janeiro

Ribeiro LCQ, Junior OAS (2017) Neoliberalization and mega-events: the transition of Rio de Janeiro's hybrid urban order. J Urban Aff 39(7):909–923

Riberio LCQ, Rojas N (2014) Rio de Janeiro in the context of Megaevents: the hegemony of a development model against the metropolis. EURA/UAA city futures conference - Cities as strategic places and players in a globalized world, Paris

Robinson J (2008) Global and world cities: a view from off the map. Int J Urban Reg Res 26(3):531–554

Sassen S (1991) The global city: New York, London, Tokyo. Princeton University Press, Princeton

Schausteck de Almeida B, Graeff Bastos B (2016) Displacement and gentrification in the 'City of Exception': Rio de Janeiro Towards the 2016 Olympic Games. Bull J Sport Sci Phys Educ 70:54–60

Taylor PJ (2010) Specification of the world city network. Geogr Anal 33(2):181–194

Teixeira OS (2015) A região metropolitana do Rio de Janeiro na atualidade: recuperação econômica e reestruturação espacial. Confins (25)

Trujillo JL, Parrilla J (2016) Redefining global cities. the seven types of global metro economies. Global cities initiative, The Brookings Institution, Metropolitan Policy Program

Vannuchi L, Van Criekingen M (2015) Transforming Rio de Janeiro for the olympics: another path to accumulation by dispossession? J Urban Res 7

Vicino TJ, Fahlberg A (2017) The politics of contested urban space: the 2013 protest movement in Brazil. J Urban Aff 39:1001–1016

Chapter 3
Public–Private Partnerships in the Context of Mega Events

Carolina Pereira dos Santos

Abstract The 2016 Olympic Games hosted in Rio City was used as pretext for the implementation of a set of urban projects in the city, some of them carried out through public–private partnerships (PPP). The case of the Olympic Park draws attention not only because of the question of the territory, but also because of the countless promises of the legacy that were not fulfilled after the Olympic Games. In addition, there were removals of the families that lived in the territory and had to give space to the implementation of the Olympic project. The arenas to which no investment has been addressed, equipment that has been abandoned, schools that have not been built, all are examples of the disregard for this legacy.

Keywords Olympic · Private–public partnership · Legacy

3.1 Introduction

The 2016 Olympic Games, hosted by the city of Rio de Janeiro, served as a pretext for the implementation of a set of urban projects in the city, some of them carried out through Public–Private Partnerships (PPPs). The PPPs of the Olympic Park, of Porto Maravilha and the Light Rail Vehicle (LRV) were implemented based on a discourse that presented them as an alternative to bring private resources to public infrastructures and to allow the allocation of risks between the public and private partners. However, the analysis shows that there was apparently no risk allocation or private funding for the PPPs, both widely taken by the public partner.

Regarding territory, what has been verified in the PPPs of Rio is that the investments were made only in two regions of the city: Barra da Tijuca and the Port Region. Both in real estate and speculative terms, these two regions have been expanding, clearly responding to the interest of the real estate market. In other words, with

C. P. dos Santos (✉)
Institute of Urban and Regional Planing, Federal University of Rio de Janeiro, Rio de Janeiro, Brazil
e-mail: santosdacarol@gmail.com

© The Editor(s) (if applicable) and The Author(s), under exclusive license to Springer Nature Switzerland AG 2020
L. C. de Queiroz Ribeiro and F. Bignami (eds.), *The Legacy of Mega Events*,
The Latin American Studies Book Series,
https://doi.org/10.1007/978-3-030-55053-0_3

the discourse of the city improvements triggering the idea of a social legacy, there was the legitimization of projects of urban reforms and restructuring. Such projects benefitted the speculative capital but there was no broad social debate about their implementation, and they did not consider the interests of the population, especially the popular classes. The case of the Olympic Park draws attention not only because of the question of the territory, but also because of the countless promises of legacy that were not fulfilled after the Games. In addition, there were removals of the families that lived in the territory and had to give space to the implementation of the Olympic project.

The arenas to which no investment has been addressed, equipment that has been abandoned, schools that have not been built, all are examples of the disregard for this legacy.

To analyze these questions, this article is divided into a few topics. First, a brief description of the Federal Law that regulates PPPs in Brazil and their dissemination in the Brazilian territory. Second, a small analysis of the neoliberal project in the context of mega events. Third, a contextualization of PPPs in relation to the mega-sporting events that occurred in Brazil in recent years and the charms of mega events. Fourth, the urban interventions resulting from Rio 2016 Olympics and the costs of the Olympic Project. Last, we will present an analysis of the Olympic Park project, of the PPP that was accomplished and of the one that was not accomplished, and the current legacy of the venture after the Olympics.

3.2   Types of PPPs and Their Diffusion in Brazil

The *stricto* sensu Public–Private partnership (PPP) was introduced in Brazil in the early 2000s. However, the first laws on PPPs were issued by the states. Federal Law N° 11079 passed only on December 30, 2004, establishing the general rules for bidding and contracting of PPPs within the union, states, federal district, and municipalities. In this way, state laws that had been sanctioned prior to this law could only be enforced if they did not oppose the federal law.

According to Law N° 11079/04, PPP is a partnership between the public agent (federal, state, or municipal government) and the private agent to provide works and services to society, with contracts ranging from 5 to 35 years, with minimum values of R$ 20 million.[1]

PPPs are different from privatizations because despite being implemented and managed by private agents, at the end of the contract the infrastructure is transferred to the public sector, with no alienation of public assets to the private sector.

They also differ from traditional concessions in which the form of remuneration for the private partner is based on the collection of tariffs for users of the services

---

[1] According to Brito and Silveira (2005), this minimum value has the goal of creating barriers to services of low-contractual value, since PPPs have complex contracts with high costs of the transaction.

granted, while in the PPP, the partnerships can be made in two ways: through the sponsored concession or through the administrative concession. In the former, the remuneration of the private partner is made in a combination of tariffs charged to users of the services with complementation of revenues by the public sector, through regular contributions plus the taxes and charges involved. In the latter, the private partner is remunerated exclusively by the public partner.

Although the sponsored PPP is the only form of partnership in which the law allows a combination between revenues from fees collection and from resource real-location through public consideration, there is a legal loophole that allows that, in a common tariff concession contract, the private partner receives subsidies from the municipalities when supported by a specific municipal law. This is the case of the Trans Olympic BRT, which is a common concession preceded by public works,[2] which means that the municipality of Rio de Janeiro makes other transfers to the private entity through subsidies, in addition to the fare charged to the user of the service.

In relation to the PPPs Law, on December 28, 2012, Law N° 12766/12 was promulgated with amendments[3] to Law N° 11079/04 concerning the general rules for public bidding and PPPs contract within the public administration. With the aim to foster investments in infrastructure made through PPPs, the law improved mechanisms and became more attractive to private investors.

Notably from 2012, with the exception of 2010, the number of PPPs and the volume of resources involved in the projects were more constant and numerous (despite the fall in resources in recent years), according to Table 3.1.

This investment instrument was originally designed for large infrastructure works but it has increasingly been used by the Brazilian State for various types of investments in other areas, mainly linked to urban policies. In most cases, the agreements through PPPs were signed considering that the public administration would be unable, by itself, to finance the works and administer the facilities. On the other hand, this allocation of the demand risks from the private entity to the public entity seems questionable since it can allow the private partner to invest without assuming any risk, only the benefits of the venture:

> In entrepreneurial governance, the comparative advantage afforded by public-private part-
> nerships would be linked exactly to the fact that the public authorities minimize the risks of
> market activities, characterized by their essentially speculative nature (Lima and Santos Jr.
> 2015).

In Brazil, 99 *stricto* sensu PPP contracts were signed, namely 49 with the municipal government, 46 with the state, three with the Federal District, and only one with the Union. The number of municipal PPPs seems to strengthen each year, as shown by the 2016 indicators (Radar PPP), 12 PPP contracts were signed across Brazil and all of them were municipal.

---

[2] Available at: <https://www.rio.rj.gov.br/dlstatic/10112/5305003/4138534/IntroducaoaoConceitod ePPPeConcessoes.pdf >. Accessed: July 31, 2017.

[3] New alterations to PPPs law available at: <https://www.cosjuris.com/novas-alteracoes-na-lei-de-parcerias-publico-privadas>. Accessed: July 9, 2017.

Table 3.1 Number of PPPs and volume of resources (in R$ billions) per year—Brazil, 2016

Year	No of PPPs	Volume of resources (R$ Billions)
2006	3	1.6
2007	3	3.9
2008	5	3.4
2009	4	8.2
2010	12	16.8
2011	3	3.0
2012	16	17.2
2013	13	44.8
2014	17	35.4
2015	11	15.5
2016	12	7.7
Total	99	157.5

Source Radar PPP (2016) with tabulation made by the author

In view of these data, apparently PPPs in Brazil, despite the discourse, did not spread extensively in the infrastructure areas of the federal government. In the face of this low dissemination at the federal level, one should inquire to what extent the PPPs would not be an experiment that has started to spread in the country from the states and municipalities. In this scenario, may the subnational governmental levels be operating as laboratories for the experimentation and dissemination of PPPs?

3.3 Political Game and Market Solution

A possible explanation for the large diffusion of PPPs between states and municipalities could be the limit of budget debt, which restricts their possibilities of expenditure. According to Raquel Rolnik (2009), Brazilian municipalities have restricted access to credit[4] and limited revenues (which end up only covering the cost of the municipal machine), which makes municipalities dependent on voluntary transfers to construction and investment in urban infrastructure. These voluntary transfers occur through covenants between municipalities and the state and federal governments.

The urban policy of the municipalities is therefore at the mercy of a political-electoral game, where access to credit depends on the relations that the local rulers establish with the federal government and the individual capacity that each politician has to bring these resources to their respective municipalities (Rolnik 2009).

---

[4]"The municipalities have at their disposal taxes that are applied on eminently urban activities: *Imposto sobre a Propriedade Predial e Territorial Urbana* (IPTU) [Urban Real Estate Property Tax] and Impostos sobre Serviços de Qualquer Natureza (ISS) [Tax on Services of Any Nature]" (Rolnik 2009).

For Ferreira and Maricato (2002), most of the Brazilian cities are financially immobilized, mainly after the validity of the Fiscal Responsibility Law.[5] In this context, partnerships with the private sector serve as salvation for many municipal governments, regardless of whether they are conservative or progressive.

Another important point is the question of the payments that PPP contracts generate between the municipalities/states and the private partner. As these transfers are made monthly or annually, they do not encompass the total amount in only one amount, but throughout the partnership contract. In this way, the city or state indirectly increases its indebtedness limit—its income—because the annual resource is not directed only to that specific work and can be used for other public works and services. The debt of the municipality or state with PPP is divided in the long term and can reach even future administrations of governments not yet elected, depending on the deadline of the PPP contract.

Although this can be considered beneficial for the management that performs the PPP, it can generate a future indebtedness to the municipality/state itself, by bringing its public accounts to a standstill in the long term. That is, in this sense, PPP can mean a risk to the "health" budget of the municipalities and states.

An example of this long-term indebtedness of public administrations is happening in Portugal in relation to the PPPs of the roads. The financial crisis that the country was experiencing was aggravated by high payments due to PPPs contracts assumed in previous years, and without proper consideration of its future tax implication (Araújo and Silvestre 2014). According to the World Bank, Portugal made 22 PPPs agreements in the road sector in 15 years (between 1995 and 2000) and, with that, accumulated a debt of approximately 13.7 million euros.[6]

This context of regulations that generate financial asphyxiation of municipalities is fundamental for understanding the objective conditions that promote the adoption of PPPs. In fact, the argument that is defended in this article is that the adoption and diffusion of PPPs by the municipalities and states are socially constructed having as a substrate an objective, legal-regulatory dimension, and a symbolic discursive dimension, which legitimizes it.

The Brazilian State passed a Fiscal Responsibility Law that prevents the indebtedness of the public power, which limits the public budget of the state and municipalities and prevents them from acquiring loans for investments. Managements are then suffocated, depending on voluntary transfers that often do not happen. On the

---

[5]The *Lei de Responsabilidade Fiscal* [Fiscal Responsibility Law] (Supplementary Law N° 101, of 05/04/2000) establishes, under the national regime, parameters to be followed regarding the public spending of each Brazilian federative entity (states and municipalities). The budgetary constraints aim to preserve the fiscal situation of the federative entities, according to their annual balances, with the objective of guaranteeing the financial health of states and municipalities, the application of resources in the appropriate spheres, and a good management inheritance for future managers. Among its items are predicted that each increase in spending needs to come from a source of correlated funding and managers need to respect issues related to the end of each mandate, not exceeding the permitted limit and delivering healthy accounts to their successors (Tesouro Nacional 2017).

[6]Available at: <https://24.sapo.pt/article/lusa-sapo-pt_2016_05_03_886392665_banco-mundial-usa-portugal-para-exemplificar-riscos-das-ppp-em-mocambique>. Accessed: July 12, 2017.

other hand, the state opens special credit lines to the private sector and allows that through a PPP the municipalities and states have access to that credit. That is, we are clearly facing political decisions because states and municipalities are without financial alternatives to maintain their governability, and regardless of the party or direction of that government, PPP ends up being considered the only, or the most viable, possibility of increasing that governability. It is a context-enforced strategy. That is, there is a political construct that imposes this alternative, or rather, that does not present other alternatives, making PPP a unique way to governability.

In this way, the municipalities are almost forced to accept this logic of market because there is a political construction that subordinates the municipality to the market. It is clear that this subordination to the market is also legitimated by whole rhetoric in favor of the PPPs, based on their effectiveness and efficiency compared with exclusively public management and the benefit of allocating risks between the private agent and the public, for example.

3.4 Neoliberal Project and the Use of PPPs

For Harvey (1996), the search for resources, jobs, and capital generates interurban competition, and so even progressive or governments of a socialist nature end up practicing the capitalist games they often criticize in search of investments for their cities.

The contradiction would precisely be in the fact that although PPP is a mechanism for securing investments for cities, they do not make the modifications and improvements that cities actually need because this new urban entrepreneurship ultimately benefits specific and speculative ventures, not the whole of the population and territory.

PPPs would be a neoliberal strategy built by the state market to cope with contemporary economic crises. Capital needs to exclude to expand (and vice versa) and for this, it uses mechanisms present in urban regimes, such as PPPs (Hackworth 2007).

By using the fallacy of the private sector efficiency and the use of private resources for public investments, PPPs are legitimated, while guiding a set of strategic actions of the State in the territorial and institutional dimension that privilege the private in face of the public, increasing or reinforcing social inequality.

3.5   PPPs in the Context of Mega-Sporting Events

In cases of mega-sporting events, the logic of market is exalted, driven by neoliberal policies. The intention was precise that the mega events that occur in the country would serve as a pretext to boost investments through PPPs.[7]

The 2014 World Cup held in Brazil strengthened these perspectives. This occurred firstly by the competition between municipalities in the country to become the head-quarters[8] or subheadquarters[9] of the event. Second, by the establishment of several PPPs in infrastructure works of the stadiums. Of the 12 works for the construction and management of stadiums/arenas, 5 of them were made through PPPs.[10]

In the case of the Olympics, the use of PPPs for the implementation and management of infrastructure and service works linked to this mega event is quite evident since the city of Rio de Janeiro had not used PPP before the mega event for any project involving the municipality.

PPPs in the city of Rio de Janeiro involve the mobility system, light rail vehicle (LRV), the urban renewal of the port area (PPP Porto Maravilha), and the construction of sports equipment (PPP of the Olympic Park), central goals of this study. But before specifically entering this discussion, it is worth reflecting on the context in which the PPPs model spreads, the Olympic Games.

In the context of these PPP contracts linked to the Olympic Project, it is possible to perceive the attempt to disseminate this model in the state of Rio de Janeiro for various sectors such as health, education, basic sanitation, among others.[11]

The PPPs model in Brazil grants the private partner, in the form of concession with total or partial public subsidies, the responsibility for the construction, financing, maintenance, and operation of assets that will subsequently be transferred to the public power. In this way, the allocations of resources are decided by the private sector and not by the public, that is, what was previously considered a public service, within the reach of all, becomes a commodity, managed by a business logic, with profit as the goal and not the welfare of citizens (Lima and Santos Jr. 2015).

Another important aspect of urban entrepreneurship would be the prevalence of investments in certain specific areas of the city over the whole of the territory. These specific areas where the projects are implemented are attractive areas for the market. These specific investments, as much as they bring some benefit to the territory, increase the risks of socioterritorial inequalities. In addition, these specific projects,

---

[7] Available at: <https://www.azevedosette.com.br/pt/noticias/copa_e_olimpiadas_trarao_de_volta_parceria_publico-privada/2193>. Accessed: Aug. 1, 2017.

[8] Available at: <https://esportes.estadao.com.br/noticias/futebol,cidades-disputam-vaga-para-ser-sede-da-copa-2014,379121>. Accessed: Aug. 1, 2017.

[9] Available at: <https://www.portal2014.org.br/noticias/237/SETE+CIDADES+GAUCHAS+SON HAM+EM+SE+TORNAR+SUBSEDES+NA+COPA+2014.html>. Accessed: Aug. 1, 2017.

[10] State PPPs.

[11] In addition, another large PPP has been studied for a large urban area of the city of Rio de Janeiro, called *Plano de Estruturação Urbana das Vargens* [The Floodplains Urban Structuring Plan], known as PEU *Vargens*. If it is carried out, it will be the largest PPP contract in Brazil.

with a small number of contractors carrying out such projects, can bring power coalitions with the ability to control the public sphere and decisions regarding public resources according to their own private interests, as highlighted by the Popular Committee (2015).

Thus, the diffusion of the PPPs model in the context of mega-sporting events in Brazil may also mean the beginning of a process of dissemination of the neoliberal model of governance for the other Brazilian metropolises (Lima and Santos Jr. 2015).

Mega-sporting events have a very large symbolic dimension, capable of attracting great attention worldwide. According to this, the organization of a mega-sporting event would bring opportunities for cities to stimulate economic growth, attract private investments, and design their image on the international stage.

The 1992 Barcelona Olympic Games would be an example of this, as they would have brought the renovation of degraded areas and the projection of the city on the international stage. However, the case of Barcelona, besides appearing to be more an exception than a rule, is also contested by some authors (Capel, Clarós, Naya and Recio 2010).

There are also several authors who analyze the sociospatial impacts and human rights violations associated with mega-sporting events in the most different cities where they are carried out, for example, the analyses on the Montreal Olympics (Le Bel 2011) and the World Cup in South Africa (Ley et al. 2011). In addition, the analyses show that mega-sporting events have also served to spread neoliberal principles and reforms, including public–private partnerships (Eick 2015; Boykoff 2013; Hackworth 2007).

In the case of Rio de Janeiro, the evaluation around the alleged benefits of the 2016 Olympics deserves caution. The social and economic impacts associated with this mega event were much criticized by several authors and social organizations (Mascarenhas 2016; Lima and Santos Junior 2015; Comitê da Copa e das Olimpíadas do Rio de Janeiro 2015) indicating that the interventions favored some regions of the city and certain economic interests. That is, with the discourse of improvements to the city that triggered the idea of a social legacy, the mega-sporting events legitimated projects of renovation and urban restructuring without broad social debate and without attending the interests of the population, especially those of the popular classes.

According to a complaint from the Comitê Popular (2015):

> The coalition of political forces, added to the interests of large contractors, accelerated the 'social cleansing' of the city's valued areas, and of peripheral areas, converted into new profitable fronts for middle-class and high-income enterprises.[…] It is a policy of relocation of the poor in the city to the service of real estate interests and business opportunities, accompanied by violent and illegal actions (p. 19).

This premise seems to be confirmed when we analyze the legates left by the Rio de Janeiro Olympics. As regards social legacy, in addition to the lack of investments in social interest housing, for example, countless promises in this area were not fulfilled, such as the transformation of the Olympic Park Arenas into schools, the use of arenas for the practice of high-performance sport for Olympic athletes, or for the population recreational use, as is the case with the Deodoro complex.

In addition, it is necessary to consider the high public expenditures that concern municipal and state debt and may impact the future of public accounts.

3.5.1 Urban Interventions and the Olympic Project Costs

The candidature of the city of Rio de Janeiro to host the 2016 Olympic and Paralympic Games was presented in 2009 and the initial estimate of costs was budgeted at R$ 28.8 billion. The costs presented (September 2016) totalized about R$ 39.1 billion. Given that the 2014 World Cup cost approximately R$ 27 billion, according to the federal government,[12] the Olympics cost about R$ 11 billion more.

However, the Popular Committee denounces that the budget cost would not be in those proportions since the public power omitted some public expenditures related to the mega event, as the construction of temporary benches for Stadium Nilton Santos (Engenhão); purchase of furniture for the Athletes Village and Media Center, cost of bodies created for the Games and indemnity of the Vila Autódromo residents, items that when added accounted for about R$ 409 million to public safes. In addition, there is an omission of public benefits linked to the PPPs of the Olympic Park and Porto Maravilha. In the case of Porto Maravilha, for example, the contract provides for the monthly public consideration of R$ 10 million, for 15 years, paid in cash, on land or in *Certificados do Potencial Adicional de Construção* (CEPACs) [Additional Construction Potential Certificates]: Clause Sixth—Public consideration, item 6.1.1.[13] In the case of the Olympic Park, the PPP contract provides for the public consideration of R$ 528 million, paid in installments over 15 years, and another land of R$ 800,000 sq. ft., located in the area where the park was built. In the same way, tax exemptions and waivers linked to several of these interventions also have not entered the Olympic budget. So, according to the dossier prepared by the Popular Committee, the Olympics costs, in addition to being higher than those disclosed, would have a public consideration far superior to private spending.

In addition, the dossier also questions the concentration of the contracts of the construction works that were carried out with some contractors (Odebrecht, Andrade Gutierrez, Carioca Engenharia, Carvalho Hosken, Queiroz Galvão, OAS, Inverpar and CCR).

From this analysis, the dossier concludes that, first, there is a violation of the right to information and to the transparency of public management, in which the City Hall, omitting information, spreads the idea that public spending would have been lower than the expenditures in the preparation for the 2016 Olympics. But mainly through PPPs, the Olympics expresses the transfer of public resources to private groups, subordinating the public interest to the logic of the market. In short, the Committee

---

[12] Available at: <https://www.portaldatransparencia.gov.br/copa2014/home.seam>. Accessed: Sept. 8, 2016.

[13] Available at: <https://138.97.105.70/conteudo/contratos/EDITAL%20PPP%20E%20ANEXOS.zip>. Accessed: July 7, 2017.

denounces that the Olympics are an instrument used by the coalition of power to promote its neoliberal project in the city of Rio de Janeiro.

3.5.1.1 The Olympic Park Project

The Olympic Park is a structure of 1,180,000 m^2, built to house sports arenas and auxiliary structures for hosting the 2016 Olympic and Paralympic Games. During the Olympics, it housed the Carioca Arenas 1, 2, and 3, the Olympic Tennis Center, the Velodrome, the Rio Olympic Arena, the Maria Lenk Aquatic Park, the Future Arena, the Olympic Aquatic Stadium, two hotels, the International Broadcasting Center, and the Main Media Center.

The Olympic Park is located in the Barra da Tijuca region, an area of great real estate and speculative expansion in the city of Rio de Janeiro (Fig. 3.1). It was built on the site where previously was housed the Nelson Piquet International Autodrome and involved the controversial removal of the Vila do Autódromo community.[14]

Fig. 3.1 Overview of the Olympic Park equipment (March 2016). In the figure, we can see Hotel, Main Media Center, Velodrome, Rio Arena, Maria Lenk Park, International Broadcasting Center, Tennis Center, Carioca Arena 3, Carioca Arena 2, Carioca Arena 1, Aquatic Stadium and Future Arena. *Source* Photo from Portal of Rio Mais Concessionaire with indications, elaborated by the author

---

[14]The community neighboring the former Jacarepaguá Autodrome went through the process of removal due to the construction of the apparatus for the Olympic Park. Of the 500 families who

3.6 The Olympic Park PPPs

Theresearch on the Olympic Park PPP revealed that, in fact, the city of Rio de Janeiro divided the concessions of the Olympic Park in two stages.[15] This fact was not transparently disclosed at any moment, either by the City Hall or by the press.

One concession was established through a PPP for the implementation of the construction works and services for the 2016 Olympics and the maintenance of some equipment and use of part of the land, and another concession would be established for the maintenance, transformation, and operation of the venues that housed the sporting equipment.

The first concession was held in April 2012, from a public–private partnership agreement between the municipality of Rio de Janeiro and the Concessionaire Rio Mais S.A., formed by the companies Construtora Odebrecht (33.4%), Andrade Gutierrez (33.3%) and Carvalho Hosken (33.3%). It is an administrative concession lasting 15 years and the overall value of the contract was R$ 1.35 billion divided into: (i) R$ 1.1 billion for construction works; (ii) R$ 147 million for the services envisaged; (iii) R$ 100 million for consulting services. After completion of the Games, it was the City Hall responsibility to inject more R$ 150 million to assist in the maintenance of the equipment involved.

In relation to the second concession, there were numerous attempts by the City Hall to carry out the public bidding, the last being held in June 2017.[16] Without success due to lack of interest of the private initiative, it was up to the city to manage the sporting equipment, while the maintenance services of the public areas of the Olympic Park were under the responsibility of the Concessionaire Rio Mais.

According to the Institute of Architects of Brazil (IAB-RJ), the winning project[17] of the Olympic Park predicted an evolution of the spaces (Fig. 3.2), transforming the equipment and other areas into a permanent legacy to the city. Among what would be the legacy would be schools, offices, leisure areas, and areas for sports practice. For the project's authors, hosting the Olympic Games and building the Olympic Park would be a unique opportunity to create "[…] a deep and enduring urban and sporting legacy to Barra, Rio and Brazil" (Gusmão and Hanway 2011).

> Rio emerged as a city of global status. The city is in the midst of its largest programme of urban and community changes seen in decades and the successful execution of the Barra Olympic Park will be an inspiring contribution to this urban renaissance. The Olympic Games event will be an important catalyst and will contribute to the development of Barra as a thriving part of Rio – valuing the sports and healthy lifestyles (p. 1).

---

inhabited the site, only 20 remained, after a process of many struggles. Residents who have been removed still try to fight for their rights. Available at: https://www.ihu.unisinos.br/78-noticias/565192-removidos-pelo-parque-olimpico-lutam-por-compensacao-mais-justa. Accessed: Aug. 28, 2018.

[15] Available at: <https://epoca.globo.com/vida/esporte/noticia/2016/08/o-preco-do-legado-do-parque-olimpico-da-barra.html>. Accessed: July 20, 2017.

[16] Available at: <https://www.rio.rj.gov.br/web/guest/exibeconteudo?id=7135102>. Accessed: July 7, 2017.

[17] Available at: <https://www.iab.org.br/projetos/1o-lugar-concurso-parque-olimpico-um-protagonista-global>. Accessed: Aug. 10, 2017.

Fig. 3.2 The Olympic Park development project: during the Games (2016), transformation of the equipment (2018) and legacy (2030). *Source* Institute of Architects of Brazil Rio de Janeiro (IAB-RJ)

3.7 The Legacy of the Olympic Park

The desired legacy is far from leaving the projected spreadsheets because several types of equipment are abandoned or in a "push-push game" to know who will administer them since the concession for their maintenance has not been defined yet. In this scenario, the legacy of the Olympic Park remains undefined.

The Ministry of Sports took control of five sporting equipment in December 2016: Tennis Center, Velodrome, Arenas 1 and 2, and the Outdoor Sand Court, the latter being built only after the 2016 Olympic Games.[18] For 2018, the Ministry of Sports predicts an expenditure of R$ 35 million with the Olympic Park, namely R$ 8 million in investments and R$ 27 million in operating costs.

The Media Complex and the Hotel, although they were built to be used during the Olympics, are private equipment. The Maria Lenk Aquatic Park has been ceded from 2008 to the *Comitê Olímpico Brasileiro* (COB) [Brazilian Olympic Committee]. The Rio Arena, which had its name changed to Jeunesse Arena, is administered by the private initiative and had a renewed concession by the City Hall in 2016.

The Olympic Park Aquatic Center, whose installation will be disassembled by the City Hall—the government body responsible for this stage, still awaits the financial transfer of the federal government to launch the public notice. The two swimming pools—both the one with three meters and the one for heating—were transported by the Army to the Fort of São João in the neighborhood of Urca. One was donated for use in the Army's Physical Training Center and the other is guarded (deteriorating) in the fort until its reallocation is decided.

The project of the *Arena do Futuro* [Future Arena] included the transformation of the equipment into four municipal schools, but the Municipal Secretariat of Education signaled that the priority was to carry out the maintenance of the existing ones, in a precarious state. Meanwhile, the current mayor of the city of Rio, Marcelo Crivella, signed a fairly controversial agreement[19] with the city of Duque de Caxias for the

---

[18]The Outdoor Sand Court was provided by the Brazilian Volleyball Confederation (CBV).

[19]Available at: <https://oglobo.globo.com/rio/estrutura-da-arena-do-futuro-sera-usada-para-constr uir-quatro-escolas-em-caxias-22525627>. Accessed: Aug. 25, 2018.

donation of the structure of the Arena and its transformation into four schools for this other municipality.

There is also the possibility of an agreement for BNDES to help finance the management of the Olympic Park through a loan to a private partner. The equipment would be included in the *Programa de Parceria de Investimento* (PPI) [Investment Partnership Program] of the public bank.[20]

The PPP of the Olympic Park promised to be an alternative to minimize costs of the Olympic Games by the public entities as they would share those costs with private entities. It also promised to leave a permanent legacy[21] to the city of Rio de Janeiro with sports equipment for the use of high-performance athletes, for the people and for social events, in addition to the construction of public schools.

However, the private–public partnership only predicted the construction of the sporting and nonsporting equipment of the site and its maintenance during the games. All the legacy and maintenance of the sports equipment depended on a new stage, that is, a new PPP, which was never carried out due to the lack of interest from the private partners.

In 2017, 1 year after the event, what was seen were abandoned arenas and unfulfilled legacies. After successive failures of bidding processes for private partnerships of administration and maintenance of the Olympic Park, and to avoid a large "white elephant," the municipal government of the City of Rio de Janeiro made an agreement with the Ministry of Sports for the maintenance of part of the Olympic Park equipment. Therefore, more public money was injected so that such "legacy" would happen but not in its entirety since several equipment remain "abandoned."

Anyway, even the arenas that are in use have an unknown future, since the *Autoridade de Governança do Legado Olímpico* (AGLO) [Olympic Legacy Governing Authority], the Ministry of Sports authority that administers such sporting equipment, will be probably abolished in June 2019.[22]

References

Araújo JFFE, Silvestre HC (2014) As parcerias público-privadas para o desenvolvimento de infraestrutura rodoviária: experiência recente em Portugal. Revista de Administração Pública 48 (3):571–593

Brito BMB, Silveira HP (2005) Parceria Público-Privada: compreendendo o modelo brasileiro. Revista do Serviço Público. Brasília

Boykoff J (2013) Celebration capitalism and the Sochi 2014 winter olympics. *Olympika*. Int J Olympic Stud XXII

---

[20] Available at: <https://agenciabrasil.ebc.com.br/geral/noticia/2017-06/ministro-do-esporte-esp era-que-parque-olimpico-seja-incluido-no-ppi>. Accessed: June 19, 2017.

[21] It is worth remembering that no social interest housing was designed for the site as a legacy after the Olympic Games. In addition, the construction of the Olympic Park involved the removal of the Vila Autódromo community.

[22] "AGLO should be dissolved on June 30, 2019, of after the completion of its duties", in https:// aglo.gov.br/quem-somos-2/#1511807709255-d464c15f-252e. Accessed: Aug. 27, 2018.

Capel H (2010) Los Juegos Olímpicos, entre el urbanismo, el marketing y los consensos sociales. In Carrer 1 17, Dossier 18. Julho

Comitê Popular da Copa e das Olimpíadas do Rio de Janeiro (2015) Dossiê Megaeventos e Violação dos Direitos Humanos no Rio de Janeiro. Olimpíadas Rio 2016, os Jogos da Exclusão. Rio de Janeiro, novembro de 2015

Eick V (2015) Aumentando os lucros (com sangue): COI E FIFA na neoliberalização global. Translated by: Daphne Costa Besen. In: Orlando Alves dos Santos Júnior, Christopher Gaffney, Luiz Cesar de Queiroz Ribeiro (eds) Brasil: os impactos da Copa do Mundo 2014 e das Olimpíadas 2016, 1. ed., Rio de Janeiro: E-papers

Ferreira, JSW, Maricato E (2002) Operação Urbana Consorciada: diversificação urbanística participativa ou aprofundamento da desigualdade? In: Estatuto da Cidade e Reforma Urbana: novas perspectivas para as cidades brasileiras. Letícia Marques Osório (org). Porto Alegre/ São Paulo

Gusmão D Hanway W (2011) Um protagonista global. http://www.iab.org.br/projetos/1o-lugar-con curso-parque-olimpico-um-protagonista-global. Acessed 10 August 2017

Hackworth J (2007) The Neoliberal City: governance, ideology, and development in American urbanism. Cornell University Press, New York

Harvey D (1996) Do gerenciamento ao empresariamento: a transformação da administração urbana no capitalismo tardio. Espaço E Debates XVI(39):48–64

Le Bel PM (2011) Os Jogos Olímpicos podem não ter fim: algumas advertências sobre o "legado" olímpico à luz da experiência de Montreal. Revista Eletrônica E-Metropolis n° 06, setembro

Ley A, Haferburg C, Steinbrink M (2001) Festivalisation and urban renewal in the global south: socio-spatial consequences of the 2010 FIFA world cup. S Afr Geogr J 93(1):15–28

Lima CGR, Santos Junior OA (2015) Impactos econômicos dos megaeventos no Brasil: investimento público, participação privada e difusão do empreendedorismo urbano neoliberal. In: dos Santos Júnior OA, Gaffney C, de Queiroz Ribeiro LC (eds) Brasil: os impactos da Copa do Mundo 2014 e das Olimpíadas 2016, 1. ed., E-papers, Rio de Janeiro

Mascarenhas G (2016) Rio de Janeiro 2016: a cidade em movimento. Revista USP, São Paulo, n. 108, 49–56, janeiro/fevereiro/março 2016

Ministério do Planejamento. Parcerias Público-Privadas. 2005. https://www.planejamento.gov.br/ ppp/index.htm. Accessed Oct 2016

RADAR PPP. PPP Summit 2016: Rumo aos 100 contratos. https://www.radarppp.com/biblioteca/. Accessed 5 May 2017

RADAR PPP. As Parcerias Público-Privadas no Ano de 2016. https://www.radarppp.com/biblio teca/. Accessed 5 May 2017

Ribeiro LCQ, Santos Junior OA (2015). Governança empreendedorista e megaeventos esportivos: reflexões em torno da experiência brasileira. In: dos Santos Júnior OA, Gaffney C, de Queiroz Ribeiro LC (eds) Brasil: os impactos da Copa do Mundo 2014 e das Olimpíadas 2016, 1. ed., E-papers, Rio de Janeiro

Rolnik R (2009) Democracia no fio da navalha: limites e possibilidades para a implementação de uma agenda de reforma urbana no Brasil. Revista Brasileira de Estudos Urbanos e Regionais 11(2):31–50, novembro

Tesouro Nacional. Lei de Responsabilidade Fiscal. https://www.tesouro.fazenda.gov.br/pt_PT/lei-de-responsabilidade-fiscal. Accessed 10 May 2017

Vainer CB (2000). Pátria, empresa e mercadoria. Notas sobre a estratégia discursiva do Planejamento Estratégico Urbano. In: Arantes O, Vainer C, Maricato E (eds) A cidade do pensamento único. Vozes, Desmanchando Consensos. Petrópolis (RJ)

Chapter 4
A Mega Event Called Official Carnival: City, Culture, and Party for the Market

Fernanda Amim Sampaio Machado

Abstract In Rio de Janeiro, the neoliberal project of the city, designed in the 1990s and effectively implemented since 2009, is potentialized with the implementation of mega events that triggered the emergence of other dynamics within the urban space not only from physical and environmental transformations, but also from legal, socio-economic, cultural, and symbolic ones. It is noteworthy that part of the transformations occurred in the cultural and artistic field where the most sensitive aspect revolves around the street carnival which acquires, from that moment on, the structural and institutional contours of another mega event leading, consequently, beyond the mercantile treatment of the Brazilian festival, to the commodification and privatization of the public spaces of the city. The mechanisms adopted by the government to organize and implement Carnival followed the same format of the other urban intervention projects of this new city model. Thus, starting from the *transformation of the street carnival into a mega event*, this article proposes a reflection on the following question: what did remain in terms of culture in the city of Rio de Janeiro in the end? What is the impact of urban restructuring on the city's cultural life?

Keywords Street carnival · Neoliberal urban entrepreneurship · Resistance movements

4.1 Introduction

The urban restructuring carried out to produce the Olympic City was developed from interventions of various types and in several areas of the city. Part of these transformations occurred in the cultural and artistic field, in which the most sensitive aspect revolves around the *street carnival*, which became, from that moment onwards, the object of interest both of the municipal public authorities and of various agents

F. A. S. Machado (✉)
Institute of Urban and Regional Planing, Federal University of Rio de Janeiro, Observatório das Metrópoles, Rio de Janeiro, Brazil
e-mail: fernandaasm@gmail.com

© The Editor(s) (if applicable) and The Author(s), under exclusive license to Springer Nature Switzerland AG 2020
L. C. de Queiroz Ribeiro and F. Bignami (eds.), *The Legacy of Mega Events*, The Latin American Studies Book Series,
https://doi.org/10.1007/978-3-030-55053-0_4

connected to the tourism and entertainment sectors. In this context, the street carnival acquires the structural and institutional contours of another mega event, giving rise to conflicts that point to the appropriation of the party by the market.

Thepresence of Carnival in the history of Rio de Janeiro is not new. Since the times of the "*entrudo,*[1]" carnival has been played and experienced by various social actors who, organized in different formats such as "*blocos,*" "*cordões,*" "*ranchos,*" "*corsos,*" and samba schools, have been appropriating and occupying the public spaces of the city. Since the early 2000s, a movement of revaluation of this party in the city of Rio de Janeiro has been observed, leading to the so-called street carnival boom (Herschmann 2013), with celebrations that last about two months and expand all over the city. Yet, up to 2009 there was no position of the City Hall concerning the street party. This meant that there was no infrastructure supply, logistics preparation, urban control, brewery marketing, or exclusive beer sale.

However, from 2009 onwards, a change in the dynamics of the party begins to occur, pointing to its transformation into a commodity and to its subordination to the interests of private groups connected to the tourism sector and the entertainment industry (Machado 2017a). It is important to emphasize that the reformulation of the street carnival did not begin at that time, randomly, or simply because it was attractive to the market. In addition to the mercantile interest centered on consumption during the event, the dominance over the street carnival became fundamental for the construction of a specific project for the city, called the "Olympic project." In order to transform Rio de Janeiro into the largest tourist hub in the Southern Hemisphere, the City Hall has adopted an annual calendar of events, through which the street carnival obtained the structural and institutional contours of another mega event. This is how the "Official Street Carnival of Rio de Janeiro" was born in 2010.

Although being an eminentlypopular important cultural expression, the "rein-forcement" provided by the City Hall has been developed from the performance of specific actors, linked to the private sector,[2] and no prior social or broad participation debate on the subject has been carried out. As we will see in this article, the proposed "reinforcement" is not developed by the city as a whole but has a cut both in terms of infrastructure and in terms of dissemination and recognition of the party itself.

The structural arrangement around Carnival corresponds to the specific format of a carnival-event, since the structure supposedly provided by the producing company is not geared to the city and the urban needs but to certain official blocks. In addition, during its route it has a complete connection with the centralities of the city, mainly the existing centrality (South Zone) and the renewed centrality (central area of the city and port area).

This means that, despite the commodification of the party, what is at stake is the mercantilization and the consequent privatization of the spaces of the city, specifically the spaces that are common, which belong to everybody. It is extremely fundamental

---

[1]The '*entrudo*' corresponded to a carnival manifestation of Portuguese origin which consisted basically of throwing objects to wet and dirty people during carnival days.

[2]This denotes continuity in the privatist treatment that the City Hall of Rio de Janeiro has been conferring to its goods and spaces.

to understand this point if we aim to perform effective political actions that will enable the reversal of this framework and the transformation of urban life.

The central goal of this article is to present the transformation of the street carnival into a mega event, within the landmark of the Olympic City, pointing out its relationship with the city-business model and highlighting the contradictions and conflicts generated from that model.

Considering that Carnival is recognizably the most important cultural expression of our city, and that the regulatory model that initially emerged only to organize this party has been applied to the other artistic and cultural expressions of the city throughout the year, it is important to reflect on the structural arrangements involving the creation of the "official street carnival" to understand the social failure coexistent with the "alleged" success of the event organizers.

4.2   Carnival as an Essential Element of the Global City Project

According to the City Hall of Rio de Janeiro,[3] the "reinvention of the street carnival" occurred within a specific moment of the city, linked to the emergence of the *Unidades de Polícia Pacificadora* (UPPs)[4] [Pacifying Police Units], the fall of crime rates, and the confirmation of the city as host of the Olympic Games. In other words, the carnival was "reinvented" by the public power within the urban restructuring project linked to the construction of the global Olympic City. If we think of the cultural universe in the city of Rio de Janeiro, we can begin to understand why the street carnival is undoubtedly pointed out as the cultural intervention of greater value for this city project. The official figures regarding the festivity leave no doubt.[5] In recent years during this period, the city has hosted about 1.026 billion tourists, responsible for occupying 85% of the hotel network, putting into motion, together with other millions of people who make up the local public of carnival, approximately 3 billion *reais*.

Since 2009, both the tourism sector and the cultural sector have been appointed as two large areas capable of boosting local economy. Based on integrated analyses of the goals of the two strategic plans implemented from 2009 to 2016, it was possible to verify that among the goals pointed out were: (i) the transformation of the city of Rio de Janeiro into the largest tourist hub of the Southern Hemisphere;

---

[3]Terms used by the City Hall itself in the presentation elaborated by *Riotur* for the disclosure of the 2014 street Carnival, as provided in the annex.

[4]The Pacifying Police Unit (*Unidade de Polícia Pacificadora*—UPP) was a program implemented at the end of 2008 by the government of the State of Rio de Janeiro, elaborated from the principles of proximity police. For further information, see: <https://www.upprj.com/>.

[5]Numbers were obtained according to the information provided by Riotur through the following link: <https://www.rio.rj.gov.br/web/Riotur/exibeconteudo?id=5914149>. Accessed: April 3, 2017.

and (ii) the conversion of the city into an important political and cultural center in the international scenario (2009 and 2013 Strategic Plan).

Diagnoses used by the city of Rio to elaborate its "goals" and "strategic initiatives" of action have expressed that, although Rio de Janeiro is one of the main ports of entry in the country, some structural problems—such as the high rate of urban violence, infrastructure deficiency, and "absence of a promotion policy for the city"—were responsible for decreasing the interest of tourists to stay longer in the city (PECRJ 2009, p. 82). Based on this diagnosis, the "Rio Capital of Tourism" program was created, whose main goals were to increase the length of the tourist stay in the city and stimulate business tourism, raising the average expenditure in the metropolis (PECRJ 2009, p. 70).

In order to achieve these goals, several actions were implemented, such as: (i) elaboration of a plan for the promotion and marketing of the city; (ii) program for the expansion of hotel infrastructure; (iii) elaboration and promotion of an event agenda; (iv) implementation of tourist signage; and (v) improvement and training of personnel and logistics of the sector (PECRJ 2009, p. 82). It was in this context of transformation of the city of Rio de Janeiro into the largest tourist center of the Southern Hemisphere, from the implementation of "Rio Capital of Tourism" project that the "street carnival" was born to the public power. As the former supervisor of the street carnival of Riotur, Alex Martins, tells us:

> Once I made a joke saying the following: […] in fact, Carnival […] was born in 2009. Because from the perspective of the public power, it just didn't exist.[…] What we had, was a need to have a city structured to hold such a big party. It had been growing there for at least eight/ten years and was on the verge of becoming an uncontrollable thing, […] there was no strategic planning for it (Martins 2017).

It is possible to think about the importance of the street carnival for the "Rio Capital of Tourism" project (and consequently for the Olympic City project) and the motives of its "appropriation" by the government, and of the private agents linked to the tourism and entertainment sectors, from two central points: (i) construction of an event agenda to boost urban consumption; and (ii) elaboration of promotion policies for the city.

With regard to the first point—when we consider that the street carnival lasts for almost two months and expands almost across the entire city, in public spaces, free of charge and with free access to all and, if we also observe the plurality and diversity of the blocks that are part of this demonstration—it is possible to perceive the potential of the festivity not only to attract tourists, but also to increase their length of stay in the city, thus fostering a mass consumption in the urban environment. From this perspective, the street carnival reveals itself even more interesting if we have as a comparative parameter the Carnival of the Marquês de Sapucaí, popularly known as *Sambódromo* [Sambodrome].

Despite the worldwide notoriety of the SambaSchools parades,[6] the Carnival circuit of the Sambodrome has some limitations, such as: (i) maximum capacity of about 120,000 people per day of parade[7]; (ii) duration of the two-day event[8]; and (iii) entry to the event is predominantly paid with high values.[9] The combination of these factors makes the street carnival a much more interesting "asset" for the Olympic City project than the carnival of the Sambadrome. However, it is worth noting that one circuit does not exclude the other, since most of the street block parades happen in the morning and afternoon, and the Carnival of the Samba Schools takes place during the night and dawn.

In addition to the fact that the street carnival is inserted in this logic of conducting events to attract tourists and to boost consumption in the urban environment, it is also extremely important for policies addressed to the promotion of the city resulting from actions linked to the urban marketing. In this sense, it is worth to comment on the relationship between culture and strategic planning, considering that many cultural practices, legitimate and authentic, were appropriated in order to propel the neoliberal project of the city.

According to Sánchez (2010), there is a connection between the creation of images of the city and the attribution of meanings given to space. An image is effectively linked to a representation and, together with the discourses, builds part of the base of the neoliberal urban restructuring. Both discourse and image are inherent components of the city project in which "[..] the discourse produced is the bearer of representations and of the worldview of those actors involved in the project" (Sánchez 2010, p. 83). Alongside the discourses forming the social consensus necessary to legitimize the construction of the Olympic City, there are other discourses responsible for boosting the "sale" of the city and for attracting the solvent users. At the time when the discourses of crisis and security were triggered, as discussed in the previous chapter, the public authorities also triggered the cultural discourse and the entertainment discourse, intimately linked to the tourism sector.

---

[6]Felipe Ferreira, in his work *O livro de ouro do carnaval brasileiro* [The Golden Book of the Brazilian Carnival], states that "[…] the fabulous growth of Cariocan revelry, with its samba schools acquiring an ever-increasing projection, would make it continue being the model for Carnival festivals in several cities in the country. The Samba Schools parade of Rio de Janeiro little by little would become the symbol of Brazilian revelry, projecting the Brazilian Carnival as the biggest popular festival in the world" (FERREIRA, 2004, p. 352).

[7]Information retrieved from the survey made by *Rede Globo*, based on the information provided by LIESA in the 2016 carnival. For more information, see: <https://g1.globo.com/rio-de-janeiro/car naval/2016/noticia/2016/02/sambodromo-do-rio-bate-recorde-de-publico-120-mil-pessoas-por-dia.html>.

[8]Considering only the days corresponding to the Special Group Schools parade.

[9]For example, in 2017 the tickets for the grandstands [*arquibancadas*] (which are the majority) ranged between R$ 220.00 (special grandstands) and R$ 500.00 (tourist grandstands) per day of parade—although there are two popular sectors at R$ 10.00 which accommodate less than 15,000 people, and luxury suites [*camarotes*] and front boxes [*frisas*] at thousands of *reais* which accommodate reduced public. For further information see LIESA (available at: <https://liesa.globo.com/2017/por/02-liesa/02-liesa_principal.html>. Accessed: April 10, 2017).

It is noteworthy that the interweaving between culture and urban entrepreneurship can be perceived from the various mechanisms implemented with the goal of differentiating and promoting cities in order to attract, consequently, capital flow (Harvey 2005). If we consider strategic planning as an activity of "promotion" and "communication" (Arantes et al. 2013), we can realize how culture becomes an essential element for the growth of city. Among the most used strategies, we highlight: (i) constructions of public equipment (museums, cultural, and leisure centers); (ii) works signed by renowned architects; and (iii) realization of cultural events and festivals (Sánchez 2010).

It is important to highlight that these strategies were reproduced in several places, generating a homogenization of the cities that were worth these practices (Harvey 2005). The more governments used these mechanisms to leverage the "position" of the city in the global circuit, the less unique and special the respective cities became. This ended up causing a decrease in the "sale" power of the cities in the market, suppressing the "monopoly advantages" that could be extracted from these localities (Harvey 2005, p. 224). Therefore, to become successful, the project of urban restructuring "[…] must seek a social rooting" through the rescue of "[…] traces of the material culture or the imaginary of society" (Sánchez 2010, p. 86–87).

In this sense, the creation of identity marks capable of differentiating a particular city from the others would depend on the elaboration and articulation of cultural elements considered unique, distinct, and irreplaceable, such as Carnival. It is of notorious knowledge that there is a collective symbolic capital linked to the street carnival, which is precisely a result of its nature and own characteristics leading it to its unique position in the imaginary of people and, because of that, making it a powerful tool for urban marketing.

It is known that we consume the city and consume its spaces, but before that we consume the image of the city, what the city represents to us, or starts to represent to us through the oriented construction of symbols by the agents that produce the urban. An attractive, fun, free, and safe image of the city is extremely functional for the tourism sector, in addition to raising the city position on the global scale of the "spatial division of consumption" (Harvey 2005, p. 175). In this sense, the street carnival creates one of the "perfect" images to achieve this purpose.

4.3 The Carnival Reinvented by the Market: The PPP Carnival

For the mega event of the "Official Carnival" to be put into practice, it was necessary to adopt some measures, both of a structural and regulatory nature. It is noteworthy that these measures follow the same path traveled to the implementation of the other urban intervention projects of this new city model, involving not only partnerships with the private sector, as consequent tax exemption reduction of the tax burden of the companies involved in the projects, but also the enactment of specific legislation for urban regulations.

Firstly, it should be emphasized that, according to the official discourse, the option to implement a Public–Private Partnership (PPP) for provision of infrastructure and the overall logistics around the street carnival was justified based on the "complexity" of the party, which would make impossible another form of project execution than delegation to a private partner.

Generally speaking, PPP corresponds to a contractual instrument "imported" from foreign law that has been regulated in our country through Law N° 11079/2004. At the international level, the term has a greater meaning than that conferred by our domestic legislation, which restricts the PPP format to the granting of services and public works and has two specific modalities: (i) sponsored partnership; and (ii) administrative partnership.

According to Law N° 11079/04, the "[…] public-private partnership is the administrative concession contract, in the sponsored or administrative modality." The former is, "the granting of public services or public works […] when it involves, in addition to the rate charged to the users,[10] the pecuniary consideration of the public partner to the private partner"[11]; and the latter is "[…] the contract for the provision of services in which the Public Administration is the direct or indirect user, even if it involves execution of work or supply and installation of goods."[12]

In addition, to fall within the scope of Law N° 11079/04, it is necessary that PPPs contracts contain a fixed term from five to 35 years, and that the contracted amounts are at least equivalent to 20 million *reais*.[13] Moreover, it is expressly forbidden to formalize partnerships that have exclusively as object "[…] the supply of manpower, the supply and installation of equipment or the execution of public works."[14]

In the specific case of the PPP street carnival, the terms for its implementation, i.e., the "actions and specifications," understood as rules, have, since 2009, been foreseen in the so-called *Cadernos de encargos e contrapartidas* ["Notebooks of charges and pecuniary considerations"], which are equivalent to the Public Notice in this arrangement. These notebooks allegedly would contain "the minimum standards" that should be fulfilled for the realization of the street carnival, in which the winning company would be the one that offered 'more', from the amount stipulated by the City Hall.

The operation was structured based on three different parts: (i) *THE DIRECTOR*—represented by Riotur, the municipality body competent for the realization and organization of the party; (ii) *THE PROMOTER/PRODUCER*—the company or consortium of companies responsible for the production of the party; and (iii) *THE*

---

[10] According to art. 2, *caput*, of Law N°. 11079/04.

[11] According to art. 2, § 1, of Law N° 11079/04.

[12] According to art. 2, § 2, of Law N° 11079/04.

[13] According to art. 2, § 4, I and II; and art. 5, I of Law N° 11079/04.

[14] According to art. 2, § 4, III of Law N° 11079/04.

FINANCIER—the company responsible for the contract costs,[15] with goals of trademark placement for promotional purposes only ("Notebooks of charges and pecuniary considerations"). It should be noted that, in this structure, the *producer* himself directly hires the project financing by means of a particular instrument, in which the City Hall of Rio participates as its *intervener*.

The object of the main contract goes far beyond the supply of "chemical toilets" or other infrastructure items, and it is for the company, or consortium of companies, responsible for the production of the street carnival to perform the "[…] operationalization, production, design, manufacture, installation, assembly, leasing of materials and equipment, maintenance and removal of equipment and all the necessary infrastructure for the implementation of the event" (Caderno de Encargos e Contrapartidas 2016, p. 3).

In this way, it was transferred to the company, or consortium of companies, responsible for the production of the street carnival, the responsibility to operate in the control of the traffic and of sales promoters; to provide a fully functioning medical-hospital structure; to carry out urban cleaning; to elaborate the audiovisual content of the event, in addition to the decoration of public parks (Machado 2017a).

Firstly, it should be emphasized that although the streetcarnival is a festivity that is naturally and spontaneously developed across the streets of the city, appropriating the public spaces as a whole,[16] actions related to the implementation of the infrastructure supporting the party are thought to supposedly[17] "[…] attend the Street Blocks parades".[18] It is noteworthy that it is not to support any and all Street Block(s), but only those "authorized" by the government, based on the municipal legislation specially created to regulate the "event," and in the "places previously determined by Riotur" (Caderno de Encargos e Contrapartidas: Carnaval de Rua 2016, p. 4).

Thus, the infrastructure that should be thought to serve the city as a whole is restricted to specific blocks parades and in specific parts of the city. From the analysis based on the public notice, it is verified that the focus of the production of the party revolves around the Center–South Zone axis. Considering, for example, the medical-hospital structure, except for ambulances that correspond to mobile structures, the installation of medical-hospital units is provided only for the Center (one unit) and for the South Zone (two units: one in Copacabana and another in Leblon), with no

---

[15]It should be noted that, in the operations of PPPs, there is usually the figure of a financier which is different from the concessionaire. At Carnival there is a financier who is different from the producer/promoter. As much as there is no finance project in terms of structured projects to finance such ventures, usually financed by the banks, the existing logic is the same.

[16]This has even been acknowledged by the government itself which understands that the "[…] party is spontaneous, in the Cariocan style of being, spreading across the city" (Caderno de Encargos e Contrapartidas: Carnaval ee Rua 2016, p. 2).

[17]One says "supposedly" since the provided structure does not satisfactorily meet the demands and needs of the carnival audience.

[18]According to the introduction of the *Caderno de Encargos e Contrapartidas* (2016): "Thinking about the importance that the street carnival and its many blocks have, the City Hall of Rio de Janeiro, through Riotur, presents below the actions and specifications for the realization of the 2016 Street Carnival with emphasis on supporting the Street Blocks parades (p. 2).

requirement of installation in the North Zone, West Zone, Governador Island, and other sub-regions. The same happens with the decoration and the distribution of chemical toilets which during the festivity are concentrated in these centralities, in a much higher proportion than in the other parts of the city (Machado 2017a).

It is worth highlighting that the Notebooks of Charges expressly provides for two pecuniary considerations guaranteed to the financier of the event, that is, as "payment" for the contribution of money made. They are: (i) the possibility of "exhibition of the brand" only in the "advertising" artifacts[19]; and (ii) the distribution of "promotional gifts," provided they have Carnival characteristics and some usefulness after the event.

Despite being only two explicitly foreseen pecuniary considerations, there is currently a very strong debate in the city of Rio de Janeiro on the monopoly of AMBEV (through its *Cervejaria Antarctica*) in relation to the sale of beer in the public spaces of the city during the pre-carnival and carnival periods. It is noteworthy that, although there is no express provision in the "Notebooks of charges and pecuniary considerations" guaranteeing the exclusive sale of beer by the project financier, the existence of such provision emerges when one reads the Notebooks' Annex.

Considering that the street vendor is registered to act exclusively for the authorized brand (which corresponds in practice to the financier's brand), and the Municipal Guard is authorized to seize any and all foreign material whatsoever, i.e., material not provided by the certified company, the exclusive sale of certain items is guaranteed and controlled by the security agents of the municipality in the public spaces of the city. In this way, the contractual structure created by the government is contrary to the principle of free competition, established by art. 170 of the Federal Constitution which aims to protect the economy from oligopolies and monopolies that threaten the economic growth and social development.

Moreover, if we think about the very structure of the PPP chosen by the City Hall, we can see that the specific structural arrangement designed toperform the carnival does not fall under any of the legal hypotheses provided by Law N° 11079/04, a law that establishes general rules for public–private partnership bidding processes and contracts in Brazil. Considering that the municipality carries out an *annual* bidding process to choose its *partners* and that it presents its contract as *non-costly*,[20] we cannot speak of a PPP under federal law precisely because the aforementioned law requires, among other matters: (i) contracts with a minimum period of 5 years and a maximum period of 35 years; and (ii) contracts with operating values above 20 million *reais*.

---

[19]From the reading of that clause is possible to realize that there is a difference between decorative elements and publicity apparatus, and that the acquired "right" of brand exposure does not cover the decorative elements, although the advertising items must follow the "patterns" of official decorations.

[20]It is worth mentioning, for example, the publication of the contract instrument extract N. 18/100/516/2012, in the Official Gazette of the Municipality, 08/08/2012, p. 89, which contains all this summarized information.

Although not falling under the normative provisions of Law N° 11079/2004, due to the factors previously mentioned, it is possible to verify that the assumptions, the language, as well as the operating and functioning logics behind these contracts are similar to those of a PPP. That is, it is actually a public–private partnership structure. It is noteworthy that in Rio de Janeiro this "partnership" has been built, since 2010, by the integration between the City Hall of Rio (*director*), the company *Dream Factory Comunicações e Eventos Ltda.* (*producer*), and *Companhia de Bebidas das Américas*—AMBEV[21] (*financier*). Although the annual nature of the bids,[22] Dream Factory has always been the winning company. It is worth noting that, with the exception of the 2010 Carnival, in which the City Hall of Rio would have received 31 proposals, in all other years (from 2011 to 2016) the choice of the company took place because it was the sole bidder.[23]

It is important to realize that the stay of Dream Factory in the production of the street carnival in the last seven years, as well as the participation of this same company in the organization of other events of a tourism nature[24]—such as Rio Marathon, Rio Bridge Race, Mano a Mano, Árvore Bradesco Seguros, Rio Gastronomia, World Youth Day, Roda Skol, and Rock in Rio—points to a resignification of local powers (Vainer 2013) linked to the cultural and tourism sectors.

4.4   The Illegitimate Norm: The Antilaw Regulating the Carnival

In addition to the structural changes, the resignification of the street carnival by the public authorities also involved the creation of a specific legislation in order to regulate the street blocks parades, considering that for implementing this model it was necessary for the City Hall to know which block would parade, at what time and where, under penalty of neither being able to exercise its actions of urban control nor minimally equipping the spaces of the city considered vital for the party.

---

[21] AMBEV's performance occurs at the Rio de Janeiro street carnival through its specific brand of beer, "Antarctica". For more information on the company's negotiating structure, we recommend accessing <https://ri.ambev.com.br/conteudo_pt.asp?idioma=0&conta=28&tipo=43218>. Accessed: January 5, 2018.

[22] According to the publications in the Official Gazette of the Municipality, as an invitation to remove the public notice for purposes of the presentation of bidding proposals.

[23] According to the publications in the Official Gazette of the Municipality of days: Oct. 30, 2014, p. 92; Oct. 7, 2013, p. 69; Oct. 11, 2012, p. 66; and Oct. 3, 2011, p. 57.

[24] The information about the events developed in the city of Rio de Janeiro under the responsibility of the Dream Factory was collected from the company's own website. For more information, we recommend seeing: <https://dreamfactory.com.br/>. Accessed: Jan. 3, 2018.

The central axis of this regulation currently revolves around the Decree N° 32664/2010[25] of the mayor of Rio de Janeiro which specifies the norms and procedures for performing carnival block parades during the pre-Carnival and Carnival periods.[26] With this norm, a prior authorization is now required so that carnival blocks and bands can parade through the streets and occupy the public spaces of the city. The competence to grant this authorization was entrusted to the Secretariat of Tourism of Rio de Janeiro, based on the opinion given by the *Companhia de Engenharia de Tráfego do Rio de Janeiro* (CET-Rio) [Traffic Engineering Company of Rio de Janeiro] and subject to the "Nothing Opposed" of the subprefectures of the city (art. 2).

In addition, to obtain the definitive authorization the blocks had to present, together with the application for authorization, the following documents: (i) application based on the form made available in the annexes to the decree; (ii) copy of the Identity Card and the *Cadastro de Pessoas Físicas* (CPF) [Taxpayer Identification Number] of the person *responsible* for the block or band; (iii) Public Security and Civil Defense authorities, *Companhia Municipal de Limpeza Urbana* (COMLURB) [Municipal Urban Cleaning Company], and *Secretaria Municipal de Ordem Pública* (SEOP) [Municipal Secretariat of Public Order] must be aware of the request, proven by the respective protocol; and (iv) proof of compliance with the requirements made at the discretion of the subprefectures.

It is necessary to realize that by "authorizing" a particular block or band to parade in the streets of the city, with a temporal limit and in a predetermined place, the City Hall of Rio is, in practice, authorizing who can or cannot culturally occupy certain public spaces during the party. It is important to emphasize that the strategy used by the City Hall of Rio to organize the party, from an "authorization," does not find any constitutional or legal support within the current Democratic State of Law. Analyzing this regulation in the light of the 1988 Constitution, and of the City Statute, we can realize that this norm does not meet the criteria of validity of the legal system, precisely because there are express provisions by higher-ranking rules of law, in the opposite direction.

After having prohibited the use of public spaces for long periods, as well as social manifestations, the 1988 Constitution provided in art. 5, IX, the protection of the free development of certain activities, including the artistic one which can be carried out independently of license.[27] Its aim was not only to safeguard the right to cultural expression of the citizens, but also to prevent cultural matters from being object of evaluation by the public authorities in relation to their convenience and adequacy. In addition, freedom of assembly and demonstration in public spaces was also ensured,

---

[25]This decree repealed the Decree N° 30659/2009, which was the first regulation applied by the city to regulate the city during Carnival.

[26]According to the art. 1 of the mentioned Decree, "pre-Carnival period is considered the thirty days preceding the Saturday of Carnival and Carnival is the period between the Saturday of Carnival and the Sunday after the Saturday of the Champions' parade". It is on this Saturday that the Samba Schools which owned the competition for the best Samba School will parade again.

[27]According to the 1988 Constitution, art. 5, IX: "The expression of intellectual, artistic, scientific and communication activity is free, regardless of censorship or license".

irrespective of authorization, if it is held peacefully, and without frustrating another previously scheduled meeting, and the sole requirement is the prior communication of such change to the competent authority.[28]

Moreover, a question that also deserves to be emphasized is that regulation is imposed by decree, which is one of the least democratic normative figures in our legal order, precisely because it is produced by the Chief Executive and does not go through the democratic legislative process. In practice, the public power has restricted fundamental rights without any broad debate on the subject.

It is important in this case to critically think the law, reflecting on the uses of the norm and how the normative institutional processes are being used in order to protect the interests of a particular group, or economic sector, to the detriment of the population in general. That is, to the detriment of those who build and play Carnival from a perspective linked to the social reproduction itself. Thus, the regulation is constructed as illegitimate right if we understand that from a critical perspective the legitimacy is attributed and conferred from the social practices, from the street (Lyra Filho 2005; Sousa Junior 2008).

In this sense, we can affirm that the establishment of a new regulatory framework, created specifically to meet the particularities of this event and to ensure the satisfaction of the interests of private partners, allowed the "legalization" of the illegality (Vainer 2011) resulting from this project. It should be emphasized that the publication of this legislation generated a contradiction between the public power, that exercised its urban control through an authoritative figure of use of public spaces, and the population, that regarded as legitimate the occupation carried out for cultural and artistic purposes, independent of authorization or license. It should also be emphasized that this contradiction did not come free from conflicts, sometimes developed violently, between agents linked to the public power and the city's population.

4.5   Final Considerations: Popular Resistance and the Worsening of the Neoliberal Logic

The legitimacy of the structure created to "produce" the street carnival (including the normative framework) was contested by several artistic and cultural groups which, over the past few years, have been building an opposition to the model of the "carnival business." If we stop to observe the political and social actions that emerged in our city due to the restructuring of the street carnival, we can perceive the emergence of the movement linked to the so-called alternative blocks, in the sense of not being part of the official carnival, building a parallel circuit, apart from the predominantly mercantile circuit of the party (Machado 2017a).

---

[28] According to the 1988 Constitution, art. 5, XVI, all can meet peacefully, without weapons, in places open to the public, irrespective of authorization, provided that they do not frustrate another meeting previously convened for the same place, and only prior notice to the competent authority is required.

As they understand that the authorization created by the City Hall to determine who can or cannot occupy the city during Carnival is unconstitutional, several blocks decided to occupy the public spaces without participating in the authorization procedure, based on their constitutionally guaranteed right. This caused the appropriations of the spaces of the city to develop in a more spontaneous, free, and democratic way, precisely because they had been conceived and developed by the people themselves, without the direct interference of the public power.

Because they are not part of the official circuit controlled by the City Hall, the unofficial blocks end up not suffering from the impacts arising from the marketing and publicity actions undertaken by the event financier. In this way, it is possible to observe a much more colorful and authentic visual identity than in the other blocks participating in the official circuit. In the spaces occupied by these blocks, informal workers act without the need for registration, which makes it possible to sell other brands of beer, different from the official beer brand (Machado 2017a).

However, security agents linked to the state government have constantly sought to take control of the public spaces of the city during the festive period, basing their actions on the norms specifically created for this event, having generated episodes of repression and violence over the past few years, for example, the episode involving the *Tecnobloco* parade on February 13, 2016 ended up with rubber bombs and bullets in the Mauá Square area, located downtown.

Despite the constant conflicts, the movement starred by the unofficial blocks[29] obtained some considerable victories. When they consider Carnival a popular expression, and not an event, the organizers of the *unofficial blocks* created a "meeting space" open to the participation of all, a space collectively constructed without any segregation by ropes or other means. From the perspective of the right to the city, this is a powerful space not only because it operates in an opposite way to the existing capitalist production logic, but also because it favors the meeting, enabling a collective appropriation of the city.

It must be emphasized that the motivations behind the actions performed by the organizers of these blocks allow interpreting this practice as an expression of the ideal of the right to the city, since the city is claimed for collective and social use. In this context, both the claim of the free and authentic carnival, and the street, as a place of its implementation, can be read from the right to the full exercise of citizenship. Moreover, the defense of common spaces with the consequent positioning contrary to their privatization guarantees that the social interest prevails over private speculative interests, shaping the principle of the social function of the city, which comes out of the constitutional abstract level to be materialized through these practices.

It is noteworthy that, besides being powerful, this movement obtained victories that were normatively recognized throughout these years. The first victory concerns the discontinuity of the privatization process of the streets that was becoming stronger since the Carnival held in 2009; and the second victory concerns the recognition of

---

[29]For further and deeper investigation on these movements, we recommend reading the book titled *Ei você aí, me dá um dinheiro aí? Conflitos, disputas e resistências na cidade do Rio de Janeiro*, by Fernanda Amim Sampaio Machado.

this model of Carnival by the State of Rio de Janeiro as a popular expression, not as an event.

When the former Mayor Eduardo Paes (PMDB/RJ) published the first decree to regulate Carnival, in 2009, several blocks began to sell shirts, following the "*abadá*" rationale which ensured the permanence within the space of the rope, previously restricted to musicians and the production of the block. Contrary to this trend, a "movement against the *baianização* of Carnival in Rio de Janeiro" was initiated, starring in the unofficial blocks, which, despite the criticism suffered, managed to reach several social actors and even the City Hall that expressly prohibited, by means of art. 1 of the Decree of the mayor of Rio de Janeiro N° 36760/2013, the use of rope or security agents for the delimitation of "private" areas in the street blocks of the city.

In January 2016, the movement of the unofficial blocks secured its second victory, thus obtaining the formal recognition of this model of Carnival as a popular cultural expression. The recognition was made through the alteration of the State Decree of Rio de Janeiro N° 44617/2014 which regulated the procedures relating to the authorization for any events that promoted the concentration of people in the state. In fact, carnival blocks began to receive the same treatment granted to "public meetings for expression of thought," no more being treated as cultural events.

With the end of Eduardo Paes' government and, consequently, with the change in the municipality management, there was a worsening of this privatizing logic. Despite discourses referring to the possible religious nature of the current mayor's interferences, Marcelo Crivella, what is perceived is the intensification of the privatization processes of urban resources in several fields. Regarding the street Carnival, it was possible to check some "novelties" in the structure of the party and the obligations of the private partner. The main one, without any doubt, revolves around the security area.

During the official presentation of the 2018 street Carnival, Mayor Marcelo Crivella revealed that the security of the party would be developed for the first time by 3,375 private security guards of surveillance companies. In addition, it was foreseen the existence of five video-monitoring centers (with 70 cameras), with capture and storage of images in charge of the company conducting the event. In other words, public security, public spaces are being delegated to private agents, in a flagrantly illegal and unconstitutional way.

This is because, according to the *Supremo Tribunal Federal* (STF) [Supreme Federal Court], in a decision conferred on the exercise of the concentrated control of constitutionality (Direct Unconstitutionality Action N° 2827, rapporteur Minister Gilmar Mendes, 9-16-2010, P, DJE of 4-6-2011), states and municipality are unable to institute or delegate the function of public security to a different body than those bodies provided for in art. 144 of the Constitution. In other words, public security can only be exercised by the Federal Police, Federal Highway Police, Federal Railroad Police, Civil Police, Military Police, Military Fire Department, and Municipal Guard (within the municipality, as provided by the law specifically designed for its regulation).

In addition, the "*Macrofunção Carnaval Mais Legal*" was enacted by Decree Nº 44217/2018 with the goals of implementing the broad control of activities developed in Carnival, and the licensing of economic activities in the public area and events of street Carnival. Of all the points dealt with in the decree, the most noteworthy is linked to street trading.

According to art. 8, the authorizations for the street trading and the market in public area, with or without the use of fixed or movable equipment, will be granted only to the legal entities that are officially sponsoring the carnival or individuals appointed by them. Considering that within this arrangement only the sponsor can trade in the public spaces of the city, we can conclude that the monopoly practice, which is legally prohibited, has just been regulated by the municipal legislation created by Mayor Marcelo Crivella.

Within this context, the norms that should be observed and the rights that should be guaranteed are massively disregarded for the realization of an extremely lucrative event (for the sponsors), and functional (for the public power). It is possible to observe that in addition to the mercantilization of the party, what is at stake is the mercantilization and the consequent privatization of the spaces of the city, specifically the spaces that are common, spaces that belong to all. Understanding this point is extremely fundamental if we aim to perform effective political actions that enable the reversal of this framework and the transformation of urban life. If we think about what was culturally left in the city of Rio de Janeiro in the extinguishing of lights, we will find in the lights on the horizon a path that despite having a private nature, is crossed by insurgent practices able to lead to another story.

References

Ambev. https://www.ambev.com.br/. Accessed 8 Dec 2016
Arantes OBF, Vainer CB, Maricato E (2013) *A cidade do pensamento único*: desmanchando consensos, 8th edn. Vozes, Petrópolis
Bakhtin M (1987) *A cultura popular na Idade Média e no Renascimento*. São Paulo/Brasília: Ed. Hucitec/Ed. Universidade de Brasília
De Barros MTGM (2013) *Blocos*: vozes e percursos da reestruturação do Carnaval de Rua no Rio de Janeiro. Master thesis. Centro de Pesquisa e Documentação de História Contemporânea do Brasil, Fundação Getúlio Vargas, Rio de Janeiro
Batista VO (2008) Os princípios constitucionais e a microempresa na ordem econômica brasileira. *Nomos* (Fortaleza) 27:317–326
Brasil. Constituição (1988) *Constituição da República Federativa do Brasil*. Senado Federal: Centro Gráfico, Brasília/DF, 292 p
Castro DG et al (2015) O projeto olímpico da cidade do Rio de Janeiro: reflexões sobre os impactos dos megaeventos esportivos na perspectiva do direito à cidade. In: dos Santos Junior, OA, Gaffney C, de Queiroz Ribeiro LC (eds) *Brasil*: os impactos da Copa do Mundo 2014 e das Olimpíadas 2016. Rio de Janeiro: E-papers, pp 409–436
Comitê Popular Da Copa e Das Olimpiadas Do Rio de Janeiro (2015) Megaeventos e violações de direitos humanos no Rio de Janeiro. Comitê Popular da Copa e Olimpíadas do Rio de Janeiro, Rio de Janeiro. https://issuu.com/mantelli/docs/dossiecomiterio2015_issuu_0. Accessed 18 Dec 2016

Comitê Popular Da Copa e Das Olimpiadas Do Rio de Janeiro. Dossiê violações ao direito ao trabalho e ao direito à cidade dos camelôs no Rio de Janeiro. Available at: https://www.observato riodasmetropoles.net/download/Dossi%C3%AA_Camel%C3%B4s_2014.pdf. Accessed 7 Jan 2017

Dream Factory. https://dreamfactory.com.br/. Accessed 10 Dec 2016

EBC Agência Brasil. Guarda Municipal é acusada de repressão violenta ao Carnaval de Rua no Rio. https://agenciabrasil.ebc.com.br/geral/noticia/2016-02/guarda-municipal-e-acusada-de-repressao-violenta-ao-carnaval-de-rua-no-rio. Accessed 17 Feb 2017

Eduardo Paes. Programa da TV – Cidade Olímpica. https://www.youtube.com/watch?v=ToA3Za nl9L8&noredirect=1. Accessed 15 Nov 2016

Escrivão FA, De Sousa Junior JG (2016) Para um debate teórico-conceitual e político sobre os direitos humanos. Editora D'Plácido, Belo Horizonte

Fernandes R (2017). *(Interview Sebastiana) Entrevista Sebastiana.* [26 abril 2017]. Interviewer: Fernanda Amim Sampaio Machado, 2017. arquivo E-3 (45 min). In: Machado FAS (ed) *Quando a cidade encontra o Carnaval*: conflitos, resistências e construção do Direito. Master thesis in Law. Faculdade Nacional de Direito da Universidade Federal do Rio de Janeiro (UFRJ), pp 226–240

Gandra A (2018). Professor da FGV defende PPPs para incrementar Carnaval de Rua. EBC Agência Brasil, Rio de Janeiro, 4 de fevereiro 2016. https://agenciabrasil.ebc.com.br/cultura/noticia/2016-02/professor-da-fgv-defende-ppps-para-incrementar-o-carnaval-de-rua. Accessed 14 Jan 14

Harvey D (2014) *Cidades rebeldes*: do direito à cidade à revolução urbana. Martins Fontes, São Paulo

Harvey D (1996) Do gerenciamento ao empresariamento: a transformação da administração urbana no capitalismo tardio. *Espaço e Debates*, ano XVI, n. 39

Harvey D (2005) A produção capitalista do espaço. Annablume, São Paulo

Herschmann M (2013) Apontamentos sobre o crescimento do Carnaval de rua no Rio de Janeiro no início do século 21. *Intercom – RBCC*, vol 36, n. 2, São Paulo

Lyra Filho R (2005) O que é o direito. São Paulo: Livraria Brasiliense

Machado FAS (2017a) *Ei você aí, me dá um dinheiro aí?*: conflitos disputas e resistências na cidade do Rio de Janeiro. Lumen Juris, Rio de Janeiro, p 2017

Machado FAS (2017b) Quando a cidade encontra o carnaval: conflitos, resistências e construção do Direito. Master thesis in Law. Faculdade Nacional de Direito da UFRJ, 242f

Martins A (2017) *Interview Riotur.* [Mar 16 2017] Interviewer: Fernanda Amim Sampaio Machado, 2017. Arquivo E-3 (85 min). Machado FAS (2017) *Quando a cidade encontra o Carnaval*: conflitos, resistências e construção do Direito. Master thesis in Law. Faculdade Nacional de Direito da UFRJ, pp 197–216

O Globo. Foliões protestam contra a violência da Guarda Municipal. https://oglobo.globo.com/rio/carnaval/2016/folioes-protestam-contra-violencia-da-guarda-municipal-18476794#ixzz3x If9BxY5. Accessed 23 Feb 2017

O Globo. Prefeitura pede policiamento ostensivo das forças armadas durante carnaval do Rio. https://oglobo.globo.com/rio/prefeitura-pede-policiamento-ostensivo-das-forcas-armadas-durante-carnaval-do-rio-22277464. Accessed 2 Feb 2018

Peixinho MM (ed) Canen, D. (Asst. Ed.) (2010) Marco regulatório das grandes parcerias público-privadas no direito brasileiro. Lumen Juris, Rio de Janeiro

Prefeitura do Rio de Janeiro. Riotur. https://www.rio.rj.gov.br/web/riotur. Accessed 2 Feb 2018

Prefeitura do Rio de Janeiro (2009) Riotur. *Plano Estratégico da Cidade do Rio de Janeiro – Pós-2016*: o Rio mais integrado e competitivo. Rio de Janeiro

Prefeitura do Rio de Janeiro (2010) Caderno de Encargos & Contrapartidas Carnaval de Rua

Prefeitura do Rio de Janeiro (2013) Riotur. *Plano Estratégico da Cidade do Rio de Janeiro – Pós-2016*: o Rio mais integrado e competitivo. Rio de Janeiro

Prefeitura do Rio de Janeiro (2016) Caderno de Encargos & Contrapartidas Carnaval de Rua

Riotur. https://www.rio.rj.gov.br/riotur. Accessed 2 Feb 2018

Riotur. Recorde de investimento para o projeto Carnaval do Rio. https://www.pcrj.rj.gov.br/web/rio tur/exibeconteudo?id=7510332. Accessed 2 Feb 2018

Sánchez F (2010) A reinvenção das cidades para um mercado mundial. Chapecó: Editora Argos

Santos Junior OA (2015a) Governança empreendedorista: a modernização neoliberal. In: Ribeiro LC (ed) *Rio de Janeiro*: transformações na ordem urbana. Letra Capital/Observatório das Metrópoles, Rio de Janeiro

Santos Junior OA (2015b) Metropolização e megaeventos: proposições gerais em torno da Copa do Mundo 2014 e das Olimpíadas 2016 no Brasil. In: dos Santos Junior, OA; Gaffney C, de Ribeiro LCQ (eds) *Brasil*: os impactos da Copa do Mundo 2014 e das Olimpíadas 2016. Rio de Janeiro: E-papers, pp 21–40

Sousa Junior, José Geraldo de (2008) Direito como liberdade: o Direito achado na rua – experiências populares emancipatórias de criação do Direito. Tese (Doutorado), Direito, UnB. Brasília

Supremo Tribunal Federal. A Constituição e o Supremo. https://www.stf.jus.br/portal/constituicao/ artigobd.asp?item=%201359. Accessed 2 Feb 2018

Vainer CB (2013). Pátria, empresa e mercadoria: notas sobre a estratégia discursiva do planejamento estratégico urbano. In: Arantes OBF, Vainer CB, Maricato E (eds) *A cidade do pensamento único*: desmanchando consensos. 8. ed. Vozes, Petrópolis, pp 75–104

Vainer CB (2011) Cidade de exceção: reflexões a partir do Rio de Janeiro. In: ENCONTRO NACIONAL DA ANPUR, XIV. Rio de Janeiro. *Anais*. https://br.boell.org/site/defaut/files/ downloads/carlos_vainer_ippur_cidade_de_excecao_reflexoes_a_partir_do_rio_de_janeiro.pdf. Accessed 6 Jan 6 2017

Chapter 5
Institutional Analysis of the Secretariat of Federal Property (SPU): The Case of Porto Maravilha

Tuanni Rachel Borba

Abstract The Porto Maravilha project comprises the revitalization of the port area of the Rio City and has large dimensions: it's the largest Public-Private Partnership (PPP) and the largest Consortium Urban Operation (OUC) in the country. Such proportions have attracted great interest and visibility, with the project being extensively debated in the media, in the political and governmental sphere, and in the academic environment. Despite all this attention, however, few analyzes focus on the role of the federal body responsible for much of the land ownership present in the project, the Secretariat of Federal Property (SPU). The purpose of this work is to analyze the institutional trajectory (s) of the SPU, and how it was cultivated, based on the performance of its superintendence in Rio de Janeiro (SPU-RJ) linked to the Porto Maravilha project. It is assumed here that change depends on investment by the endogenous actors and bifurcation illustrates the fact that these investments, explained in terms of cultivation or neglect, are not exclusive, because the same actors cultivate or neglect certain trajectories depending on the interests at stake. This is the main hypothesis of this research: the coexistence of institutional trajectories in SPU-RJ.

Keywords Public property · Institutionalism · Port region

5.1 Introduction

The Porto Maravilha project comprises the revitalization of the port area of the city of Rio de Janeiro and has large dimensions: it is the largest *Parceria Público-Privada* (PPP) [Public-Private Partnership] and the largest *Operação Urbana Consorciada* (OUC) [Urban Partnership Operation] in the country. Such proportions have attracted great interest and visibility, and the project was extensively debated in the media, in

T. R. Borba (✉)
Institute of Urban and Regional Planing, Federal University of Rio de Janeiro, Rio de Janeiro, Brazil
e-mail: tuanniborba@gmail.com

© The Editor(s) (if applicable) and The Author(s), under exclusive license to Springer Nature Switzerland AG 2020
L. C. de Queiroz Ribeiro and F. Bignami (eds.), *The Legacy of Mega Events*,
The Latin American Studies Book Series,
https://doi.org/10.1007/978-3-030-55053-0_5

political and governmental spheres, and in the academic environment. Despite all this attention, however, little appreciation has been given to the role of the federal body responsible for much of the question of the land ownership present in the project, the *Secretaria do Patrimônio da União* (SPU) [Secretariat of Federal Property]. From a total of 5 million square meters of Porto Maravilha, 63% was under the Union's possession in 2009, year of the project launching, making the SPU a key actor for the feasibility of the venture, especially regarding its financial engineering based on land valuation for the generation of resources needed to PPP payment. The stock of public land circumscribed in the area and its consequent destination is, therefore, the starting point for the analysis of the performance of the SPU in the context of Porto Maravilha. In this sense, this work aims to conduct an institutional analysis of the SPU focusing on the *Superintendência do Rio de Janeiro* (SPU-RJ) [Superintendence of Rio de Janeiro].

From the perspective of the historical neo-institutionalism, in this work we explore the concept of path dependence both from the static point of view (continuity) and from the dynamic point of view (change). In historical neo-institutionalism, the temporal dimension in the succession of events is investigated in order to reconstruct the trajectory of the institution to identify the causal explanations of its current format. From this angle, we consolidated the explanatory capacity of this theoretical current for the social causality linked to trajectory, called "path dependence."

This is commonly characterized by the continuity of social processes that resist attempts to change due to the dynamics of increasing returns (Pierson 2000).[1] However, some authors argue that the same phenomenon may explain the change: increasing returns can become stronger for a new trajectory, which would weaken this mechanism in the previous one; and another possibility is that they generate a bifurcation, where two heterogeneous trajectories coexist (Deeg 2005). If, on the one hand, the "vocation" of historical neo-institutionalism for continuity has already been acknowledged, on the other hand, its aptitude to serve as an explanatory model of change has not yet been consolidated, especially due to the few empirical studies on the question.

The goal of this work is therefore to analyze the institutional trajectory (s-ies) of the SPU,[2] and how it was cultivated, from the performance of its superintendence in Rio de Janeiro (SPU-RJ) in the Porto Maravilha project. It is assumed here that change depends on investment by the endogenous actors and bifurcation illustrates the fact that these investments, explained in terms of cultivation or neglect, are not exclusive, because the same actors cultivate or neglect certain trajectories depending on the interests at stake. This is the main hypothesis of the research: the coexistence of institutional trajectories in the SPU-RJ.

---

[1] The occurrence of increasing returns can be described as a process of self-reinforcement or positive feedback (Pierson 2000, p. 252).

[2] Still according to Deeg (2005), the idea of path is understood here as the logic generated by the interactions of an institutional system.

5.2   The Historical Neo—Institutionalism: Continuity and Change

In the classic definition of path dependence, the costs to leave the path initially chosen are very high, which consequently reinforces the path initially chosen. This concept is related to the work of Arthur (1994), considered a milestone of the discussion in the context of economics, especially for evidencing the dynamics of increasing returns.[3]

Arthur (1994) argues that in the technology-intensive sectors the initial costs are high, but as production increases the unit cost decreases and there is a scale gain, which generates increasing returns and favors the permanence in the initial trajectory. He also adds that increasing returns occur and, consequently, the reinforcement of the trajectory occurs when four characteristics are present: (1) high fixed costs, generating great incentives to continue the investment in the already established option; (2) learning effects, promoting the accumulation of knowledge due to the operation of complex systems and higher returns; (3) coordination effects, so that the benefit enjoyed by an individual when he uses a good increases as other people also use it; and (4) adaptative expectations, produced when the individual chooses the option that has the biggest chances of being chosen by other individuals.

Pierson (2000) adapted this discussion to politics, arguing that, compared to economics, politics is a more fertile ground for the continuity of trajectories. He explains that what makes policy scope more resilient to changes than economics is the absence (or weakness) of mechanisms of competition and learning, in addition to the predisposition to short-term time horizons and the strong bias of the *status quo*. For the author, the previous steps in a certain direction tend to reinforce the path already underway, generating increasing returns (positive feedback). In this sense, the key elements of path dependence are the costs of change, as well as the question of time and sequence. Both Arthur (1994) and Pierson (2000) admit the occurrence of changes, adding that a specific alternative is not permanently "trapped," and changes usually occur as a result of external events.

These works on the continuity of social processes based on the dynamics of increasing returns are anchored on the persistence of practices that resist attempts to changes. At the same time, skepticism on the assumption that change would occur only through external shocks prompted the search for alternative explanations to change. More recently, authors affiliated with the historical neo-institutionalism (Thelen 1999; Crouch 2005; Streek and Thelen 2005; Deeg 2005; Mahoney and Thelen 2010; Fioretos et al. 2016) engaged in the construction of frameworks able to accommodate the question of institutional reproduction and change, devoting great attention to the role of internal actors.

---

[3] Authors such as David (2007) and Page (2006) argue that there is not necessarily a connection between increasing returns and path dependence. For Page (2006, p. 113), path dependence is based on negative externalities variables. For the purpose of this article, our starting point is the definition of Arthur (1994) which is the most cited reference in the debate on path dependence in the institutionalist literature (see also Thelen 1999; Pierson 2000).

Thelen (1999), one of the exponents of this group, argued that the key to understanding these phenomena is to accurately specify the reproduction mechanisms that support these institutions. In his work with James Mahoney, the author presented one of the most well-known theoretical frameworks on the occurrence of gradual institutional changes (Mahoney and Thelen 2010). Based on two variables, i.e., the institutional characteristics and the political context, they created a typology for the possible patterns of changes in each case, relating them to a specific type of agent and strategy.[4] Considering these variables, they claim that the basic characteristics of institutions contain possibilities for endogenous changes and that distributive implications of power is what inspires change.

Crouch (2005) endorsed criticism of path dependence: according to the author, the actors can produce changes from the elements of the institution itself and recombine them in a different way in due course. Faced with these attempts to combine path dependence with situations of changes in the environment, Crouch (2005) argues that it is possible to bring greater precision to the analysis by characterizing the individual as a Bayesian decider with his own Pólya urn,[5] noting that this decision-making exercise is not a strategic action model, because the individual seeks to respond to an environment that is not itself a strategic player.

Deeg (2005) also analyzed the processes of institutional change using the argument of path dependence. The author claims that this can also be used to explain institutional innovation, and, in this perspective, he proposes a "measurable" conceptualization of change to distinguish between transformations occurring within an existing path (on-path) and those that result in a new one (off-path). For Deeg, trajectory means the institutional logic generated through the interactions of a given institution or institutional system (Deeg 2005). From this conception, he states that: (1) the changes that generate new trajectories can have as source the actions undertaken by the actors that are within the institution; (2) the increasing returns may become stronger for a new path, which would weaken that mechanism in the old path, or, still, they can generate a bifurcation in which two heterogeneous paths would coexist; and (3) for the occurrence of increasing returns it is necessary that they be "cultivated" by the actors, through mobilizations in the political arena or organization of a collective action for the purpose of building coalitions, for example.

The works cited above have much contributed to make historical neo-institutionalism a theoretical framework capable of generating hypotheses, not just a theory designed "to tell stories" and, therefore, are consistent with the proposal of our research. The authors developed theoretical instruments aiming to capture the mechanisms of self-reinforcement or of change to which endogenous agents are exposed.

---

[4]Each variable has two categories (a total of four). The combination of the theoretical framework originates four types of change: displacement, overlap, drift and conversion. The same explanatory categories originate four types of agents: insurgents, parasites, subversive and opportunistic.

[5]In the model of Polya urn, used to illustrate the process of increasing returns, there are initially four balls, two red and two white. In each round a ball is selected; if a red ball is chosen, it is placed back in the urn along with another ball of the same color. If a white ball is chosen, it is replaced in the urn next to an additional white ball. In Crouch's proposal (2005), both the individual and the environment have an urn.

These guidelines on the research addressed to an institutional analysis support the proposal of this work in the sense that the researched institution (SPU-RJ), in the context of Porto Maravilha project, shows ambiguities in its trajectory. This ambiguity configures the hypothesis of different logics guiding the performance of the SPU-RJ.

Possible incongruences and intersections are part of alternative dynamics within path dependence (Orren and Skowronek 1994 apud Thelen 1999; Crouch 2005), to the extent that different institutional logics are present over time, fruits from broader or local contexts. The question of cultivation and neglect, resulting from the mobilization of actors, is crucial to understand the processes of change that do not result from external shocks. The cultivation, necessary in the early stages of a process of change, is carried out by the actors until the increasing returns of the new trajectory are strong enough to consolidate a new or parallel path—in the latter case, the old path can still generate increasing returns, causing the coexistence of trajectories.

5.3   The Secretariat of Federal Property (SPU)

The Secretariat of Federal Property (SPU), responsible for the management of property assets at the federal level, was established in 1854 and its performance always focused on revenue collection from the use of the Union's properties. As pointed out by Reschke (2010, p. 18), "The State favors the logic of earning profit with its assets, rather than prioritizing the public and social interest of these areas."

In addition to the establishment of this logic of action, internally the SPU was very aligned with the paradigm of operation of the *Ministério da Fazenda* [Ministry of Finance], admittedly supported by the rationality of bureaucrats (Loureiro et al. 1998). This rationality is revealed in centralized decision-making processes,[6] in the lack of participation[7] and social control[8]—practices present in the SPU until then.

Due to changes in the structure of the ministries, in 1999 the federal body ceased to be part of the Ministry of Finance and joined the Ministry of Planning. Four years later, in 2003, a new team took over the SPU management and the *Grupo de Trabalho Interministerial* (GTI) [Inter-Ministerial Working Group] was created[9] in order to restructure the federal body. Based on the GTI final report, there were changes in the guidelines and reformulation of the institutional mission:

---

[6]Only in 2004 a collegiate board was created with the inclusion of the regional agencies in the discussion of Strategic Planning.

[7]Only in 2009 institutional spaces were created for the discussion of the destination of the areas and buildings of the Union with the organized civil society.

[8]The dissemination of annual management reports, the standardization of the disclosed information, and the launch of a digital platform are recent actions of the government body.

[9]Comprising the Civil House, the Federal Attorney General's Office, Ministries of Planning, Social Security, Cities, Defense, Environment and Finance.

To know, ensure and guarantee that each property of the Union fulfils its socio-environmental function, in harmony with the fundraising function and in support of the strategic programs for the Nation (Secretaria do Patrimonio da União 2010, p. 7).

Another important result of this moment was the formulation of the *Política Nacional de Gestão do Patrimônio da União* (PNGPU) [National Policy for the Management of Federal Property]. According to the PNGPU, the management of properties should be geared toward the allocation of properties that meet public policies aimed at "social inclusion, environmental preservation, sustainable economic development and implementation of infrastructure" (Secretaria do Patrimonio da União 2010, p. 9). In this sense, the destination is guided by the understanding that each property of the Union needs to fulfill its social-environmental function, a function that implies the fulfillment of social demands, aiming at the reduction of regional inequalities and the basic social right to housing, in harmony with the management of natural resources, in order to ensure the preservation of the waters, forests and federal lands.

The development of the urban policy itself brought special contribution to the contours of the new trajectory announced in 2003, considering, above all, the influence of the urban reform agenda on the PNGPU. This agenda advocates the social function of property and democratic management in cities—principles consolidated from the approval of the *Estatuto da Cidade* [City Statute] (Law N° 10257 of 2001), which regulates the chapter of urban policy established in the 1988 Federal Constitution. It is the social function of the State to direct all the development and implementation of the urban policy in terms of appropriation, ownership and land use, to build a socially more balanced space (dos Santos Junior et al. 2011). On the other hand, the democratic management relates to social participation and the use of innovative management processes in the discussion and application of urban planning. Both principles are directly related and have their compliance subject to the Master Plan, where the instruments of their application will be defined in the different areas of the municipality.

The PNGPU was implemented in 2004 and some of its projects observe the principles outlined above, as well as the SPU guidelines of socio-environmental destination. One of these initiatives is the project "*Nossa Várzea: cidadania e sustentabilidade na Amazônia*" [Our Floodplain: citizenship and sustainability in the Amazon] in partnership with the *Instituto Nacional de Colonização e Reforma Agrária* (INCRA) [National Institute of Colonization and Agrarian Reform], where after the mapping of about 250,000 families that inhabit floodplain areas in Amazonia,[10] the *Termo de Autorização de Uso* (TAU) [Term of Use Agreement] was institutionalized, important land regularization instrument. The objective was to ensure ownership (enabling proof of residence and access to benefits) and the end of the exploitation of riverside families by land grabbers (*grileiros*).

Another initiative was the creation of the SPU National Working Group aiming to define criteria for the destination of the Union's properties for housing provision programs of social interest and to strengthen democratic management. This

---

[10] Areas that are submerged in much of the year and, therefore, set up a federal domain.

national group was composed of representatives of the same segments which form the *Conselho Nacional das Cidades* [National Council of Cities] and the experience was replicated to the state scope in 2009. *Grupos de Trabalho Estaduais* (GTES) [State Working Groups] are the main channel of dialogue and participation of the civil society organized in the SPU decisions and aim to know the demands of society for *Habitação de Interesse Social* (HIS) [Social Interest Housing]. In the same year the GTES were implemented, the federal government created the *Programa Minha Casa Minha Vida* (PMCMV) [My Home My Life Program], including a modality named Entities.[11] The SPU participation in the program is linked to the availability of land and it is within the GTES scope that social movements articulate their demands together with the Union.

What was shown above provides examples of initiatives signaling that the SPU is using its property for socio-environmental purposes. This investment implied the mobilization of the federal body for the approval of legal instruments and governance that enabled the implementation of these initiatives, as many of the resources needed for the implementation of the PNGPU were not foreseen in the legislation. Another notorious investment was the articulation with other governmental agencies for the construction of partnerships and constant interlocution. In addition, the urban reform agenda was important for the construction and dissemination of principles that were later incorporated by the group that took over the SPU management.

Nonetheless, the implementation of the PNGPU was also a challenge since it faced an unfavorable institutional scenario: scarce financial resources, limited technical framework and centralized structure. At the time, in virtually all states there was a *Gerência Regional do Patrimônio da União* (GRPU) [Regional Management of the Federal Property] (later called Superintendencies); however, the absence of organizational structure and the limited technical body implied decisions and processes centralized in the central body.

To circumvent this situation and subsidize the new policy implementation, the following actions were taken, namely regimental changes, claims of incremental budget, execution of a public tender for provision of technical framework, organizational restructuring, in addition to the creation of results-based management mechanisms.

Moreover, considering the history of the SPU, it was necessary that the endogenous actors incorporated the change in the logic previously applied, not only in the formal and normative sense. The previous logic, based on fundraising, would have to give space to the objective highlighted by the new policy: to allocate the assets in a strategic way, prioritizing the socio-environmental function. The operation of the two, not necessarily exclusive, would represent different ways of using federal property and, therefore, would influence the PNGPU outcomes and other correlated policies.

More than announcements of changes in speeches and documents, the specificities in the destination of the Union's property, according to each logic, are shown when

---

[11] This modality addresses the construction of housing for low-income families organized in housing cooperatives, associations and other private non-profit entities (Caixa Econômica Federal 2017).

we closely observe the performance of the SPU in specific projects that gather a large number of actors and, consequently, articulate different interests around the Union's assets. In this sense, the case study of Porto Maravilha represents an opportunity to observe the logics in operation in the Superintendence of Rio de Janeiro (SPU-RJ).

5.4 The Case of Porto Maravilha

In the course of the twentieth century, the port of Rio de Janeiro was losing its commercial importance and the economic and administrative activities were decreasing since the 1960s. In this sense, many buildings and warehouses have been gradually abandoned, including public properties, thus generating a considerable contingent of deteriorated buildings over the years.

The discussions about the revitalization of the area emerged in the 1980s, motivated especially by the centrality of the region and consequent commercial interests in an area of about 5 million square meters. For the purpose of contextualization, we present the schematization of Sarue (2016) that divided the proposals for revitalization into four main moments:

> The initial phase, marked by the first projects of revitalization of the port area led by the local private sector and marked by the coming to Brazil of Catalan consultants after the Olympics in Barcelona in 1992, with the proposal to associate the port revitalization with the Olympic Games; a second phase, corresponding to the formulation of the Porto do Rio Project, led by the city of Rio de Janeiro and the Instituto Pereira Passos; followed by a third phase, in which the debate shifts to the federal government, in a period of reformulation of the project, induced by the creation of an inter-ministerial executive group, coordinated by the Ministry of Planning that defined important conditions for the current policy; and finally, the fourth phase, when the structuring of the current project from direct negotiation occurred, already at the municipal level, with private actors interested in revitalization, culminating in the current proposal, approved in the management of the [then] current mayor, Eduardo Paes (Sarue 2016, p. 82).

Our work will focus the last phase; however, the change in the role of the federal government within these proposals stands out. Rising sharply in the third moment, from the *Programa de Reabilitação de Áreas Urbanas Centrais* (PRAUC) [Central Urban Areas Renewal Program] of the Ministry of Cities, the Federal Executive branch created a group with the participation of the city of Rio de Janeiro in 2004 to propose a new strategy to revitalize the area based on the program guidelines, including the possibility of a public consortium in such a way that the federal government could directly monitor the actions. As regards the Union lands, direct actions were envisaged:

> The report then presented by the group, the 'Conditions for use of the Union's properties' and the possibilities of 'Direct interventions of the federal government' highlighted the production of more of 20000 housing units, for various income ranges, in the available land, through resources of the FGTS, FAR [Rental Fund for Residence] and the General Budget of the Union [...] the Ministry of Cities worked with the prospect of building 2478 housing units, from lands of the Union (Sarue 2015, p. 87).

As a result of the working group, in 2006 there was the signing of a Technical Cooperation Agreement between the Union and the Municipality of Rio de Janeiro. This agreement aimed at "[...] the improvement of road and rail access, the implementation of housing developments in public properties of the Union and the transfer of public land in the port area for social uses and renewal projects" (Ministério das Cidades 2005, p. 34).

However, the mayor of Rio de Janeiro at the time, César Maia, did not attend the signing ceremony of the agreement, in a demonstration of his discontent with the proposal—motivated, according to Werneck (2016), by a dispute of the leadership of the project. This episode marked the return of the construction of the proposal to the local scope. In August 2006, the City Hall published the *Procedimento de Manifestação de Interesse* (PMI) [Expression of Interest Procedure] for conducting a study by private companies that assessed the feasibility of a concession or a public-private partnership (PPP) for intervention in the port area. The only consortium that participated in the PMI was formed by the companies OAS, Odebrecht, Carioca Cristian-Nielsen and Andrade Gutierrez (the last company left during the process). In the final report published in 2008, the consortium proposed the issuance of *Certificados de Potencial Adicional Construtivo* (CEPACs) [Additional Construction Potential Certificates] to finance the project and the availability of federal land in the form of donation (Rio de Janeiro 2008, p. 60).

Even in face of these changes that resulted in a different position for the federal government in the project, its participation and support would be indispensable for municipal aspirations due to the large amount of land under its property in the area. This collaboration eventually materialized and, to explain it, it is necessary to rescue the political, economic and international context that transformed the city of Rio de Janeiro from 2009 onwards.

5.4.1 The Construction of the Coalition Around Porto Maravilha

Eduardo Paes took over the city of Rio de Janeiro in 2009 and in February of the same year there was the consolidation of the application of the city to host the 2016 Olympic Games. In addition to the Olympics, in 2007 Brazil was announced as the host of the 2014 World Cup, and the city of Rio de Janeiro was one of the favorites to receive matches and to host other necessary structures such as the Media Center. The realization of these major sporting events has contributed to consolidate and legitimize the urban transformations desired by the government.

During the campaign, Paes expressed his intention to transform the port area into a tourist area. On the day of his victory in the municipal elections, the press highlighted:

Paes showed up to celebrate the victory alongside the governor of the state, Sérgio Cabral, to whom he dedicated his victory in the second round. Embracing the governor, Paes also thanked President Lula and said that the people of Rio will see a work of partnership between the federal government and the City Hall (Eduardo Paes é Eleito… 2008).

Sérgio Cabral, Governor of Rio de Janeiro, was affiliated with the same party of Eduardo Paes, the *Partido do Movimento Democrático Brasileiro* (PMDB) [Party of the Brazilian Democratic Movement] and was his political godfather. The PMDB also had a political alliance with the *Partido dos Trabalhadores* (PT) [Workers' Party], the party of the then-President Luiz Inácio Lula da Silva, and was negotiating the nomination of one of its members to compose the list of candidates of the PT in the presidential elections of 2010. At that moment, President Lula enjoyed a great popular approval (72% in 2009, according to Datafolha 2009) and the national and international media announced a very optimistic economic scenario for the country.

The political-party alignment between the three federative entities enabled the launching of the first stage of the revitalization works of the port area in June 2009. At the time, a cooperation protocol was signed between the three entities and the bills referred to the City Council were also presented. In October of the same year, Rio de Janeiro was confirmed as the 2016 Olympics headquarters, favoring the approval of the projects that had the official endorsement of the City Council in November 2009 with the publication of two *Leis Complementares* (LCs) [Complementary Laws].

The LC Nº. 101 provided the creation of an *Operação Urbana Consorciada* (OUC) [Urban Partnership Operation][12] and of an *Área de Especial Interesse Urbanístico* (AEIU) [Special Urban interest Area] in which the actions of the operation would be concentrated since the Master Plan of the municipality needed to be changed. On the other hand, the LC Nº 102 established the *Companhia de Desenvolvimento Urbano da Região do Porto do Rio de Janeiro* (CDURP) [Rio de Janeiro Port Region Urban Development Company], a mixed economy company controlled by the municipality, created in order to coordinate and monitor the concessions, as well as administer the assets and financial resources of the project.

These laws were mostly envisaged for the second phase of the project. The first phase, budgeted at R$ 139 million, was planned with the objective of encouraging the investors interest and counted on municipal and the Ministry of Tourism resources. The second phase, budgeted at R$ 8 billion, was made possible from a PPP with Porto Novo S/A (formed by the construction companies OAS LTDA, Norberto Odebrecht Brasil S.A. and Carioca Christiani-Nielsen Engenharia S.A.).

5.4.1.1   The Financial Engineering of Porto Maravilha

For PPPs, the financial engineering of the project defined the issuance of 6,436,722 CEPACs—securities that allow the incorporation of the right to construct above the basic coefficient of utilization. For the sale of these securities, an auction was organized in 2011 and the total amount was purchased by the *Fundo de Garantia por*

---

[12]An instrument of urban policy that aims to contribute to the recovery of degraded urban areas through interventions coordinated by the municipality, with the participation of residents and private investors (according to art. 32 of the City Statute).

Tempo de Serviço (FGTS) [Time of Service Guarantee Fund], managed by *Caixa Econômica Federal* (CEF) [Federal Savings Bank]. Due to this purchase, it became the administrator of the *Fundo de Investimento Imobiliário do Porto Maravilha* (FIIPM) [Porto Maravilha Real Estate Investment Fund] and has priority in the purchase of public land in the area.[13]

CEPACs were purchased by R$ 3.5 billion and the FIIPM pledged to pay the remainder of the urban operation in 15 years. According to the Fund, the business had a great potential for profitability, and its representatives pointed out the forecast of the financial valuation of the acquired securities:

> The nominal value of all CEPAC stock was R$ 3.5 billion, which gave an initial value of R$ 545 per unit. What we expect is the valuation of costs, something that is already happening. Today, the CEPAC unit more than doubled since we purchased it. What we have in the case of Porto is a virtuous cycle. We will finance the urban operation and monetize the Fund. We invest in revitalization and then harvest the fruits. We do not donate money, but investment, working for return and profit ("Entenda o negócio" 2013).

As a condition for the transfer of R$ 3.5 billion, the municipal government should provide, in up to three years, the land capable of absorbing 60% of the construction potential: seven priority lands met this percentage requirement. From this amount, six pertained to the Union, including two of the *Companhia Docas do Rio de Janeiro*[14] and two of the extinct *Rede Ferroviária Federal* (RFFSA) [Federal Railway Network], besides one that belonged to the *Companhia Estadual de Águas e Esgotos do Rio de Janeiro* (CEDAE) [State Water and Sewage Company of Rio de Janeiro].

The availability of the remainder of the amount necessary to cover the total of the operation was directly linked to land valuation—this is the basis of the financial arrangement headed by the CEF. As Sarue explains (2015):

> A direct implication of this arrangement is the need for *Caixa Econômica Federal* to hit a combination of investments that enable the necessary cash flow to the schedule of payment of PPP final compensation, speculating with the assets (lands and CEPACs) from the time of valuation of the real estate market and recovering the land valuation produced by the state interventions in the area, interventions that traditionally would be channeled only to the private sector, from the real estate profit. To meet these needs, *Caixa Econômica Federal* developed a series of business models which can authorize the institution to assume part of the risks of venture valuation regarding developers and their capitals through a partner relationship in which it participates with a portion of the CEPACs or land, with division of profits and risks and, as a result, the developers do not commit themselves to a high initial investment to operate the purchase of land or CEPACs. Finally, the decision on ventures in the area becomes competence of the management committee of the real estate investment fund controlled by *Caixa Econômica Federal* [the FIIPM] the municipality of Rio de Janeiro and the CDURP itself which, although participating in an advisory committee, has no formal decision-making power over the nature of the ventures (Sarue 2015, p. 89).

---

[13] The rules for the application of the FGTS resources, previously limited to housing, basic sanitation and urban infrastructure, changed so that operations linked to urban infrastructure could receive investment from the fund (Werneck 2016, p. 120).

[14] *Companhia Docas* are companies operated by the federal government to administer the ports of the country.

To further elucidate the matter, it is necessary to explain the real estate investment funds of the project that are divided into two: *Fundo de Investimento Imobiliário do Porto Maravilha* (FIIPM) [Real Estate Investment Fund of Porto Maravilha], as already mentioned, and *Fundo de Investimento Imobiliário da Região do Porto* (FIIRP) [Real Estate Investment Fund of the Port Region]. In the latter are located the CEPACs and the properties purchased by the municipality (such as the seven lands cited above), constituting its capital and having the CDURP as its sole shareholder (in the FIIPM the FGTS is the sole shareholder).[15] The transfer of CEPACs and the provision of land corresponding to 75% of these certificates[16] are responsibility of FIIRP, and CEF, when it obtained the totality of CEPACs, acquired the purchase priority of 60% of these lands (the other 15% can be negotiated by CDURP directly with the market).

Given the above, two notes are especially important for this work: (1) the release of resources to finance the project was directly linked to the provision of lands of the Union, since they formed practically the totality of the 60% of CEPACs required by the CEF; and (2) the deadline set by the CEF for the provision of lands (three years) implied fewer chances of valuing them in the face of the area's own revitalization works. This could also elicit a possible conflict of interest, since the CEF itself, although in another context of its bureaucracy, was responsible for the economic valuation of the land.[17]

The CEF legitimated the market logic by highlighting the profitability of the business, arguing that the remuneration would be in accordance with the investments made. However, it admits that it is a long-term profitability and that the real estate development of the region has its own pace and, in general, slower than that of the urban operation project. Even if it is possible to conclude that the speculative potential returned to the federal government, represented by the FGTS funds, there was a clear change of direction regarding the destination of the Union's assets. The City Statute, acting as a legal framework for encouraging social function, and the very discussion of the project in the PRAUC of the federal government, indicated the use of mechanisms for the construction of a trajectory, and consequent destination of the land focused on the socio-environmental function. However, there were changes: if at first the land was considered strategic for the promotion of housing, including housing of a social nature, from 2009 onwards they became important assets in the complex financial engineering elaborated to afford the PPP of Porto Maravilha—meeting a destination here called "pro-market."

---

[15]The constitution of two funds is explained, according to the CEF, by the need to ensure the governance of the assets (decision making would be tied to the majority shareholder).

[16]The remaining 25% are private lands.

[17]The responsible for the economic valuations was the *Gerência de Desenvolvimento Urbano* (GIDUR) [Management of Urban Development].

5.5 The Federal Property in the Context of Porto Maravilha

From the interviews conducted with the key actors in the context of the performance of the SPU-RJ in Porto Maravilha,[18] we sought to analyze how the trajectory based on the destination of the Union's assets was cultivated and what was its meaning in the project. The main common thread of this analysis is the coexistence of distinct institutional logics: the fundraising logic has never ceased to exist, but, on the other hand, and more importantly, the destination logic also bifurcated, serving distinct purposes due to the cultivation of the actors.

5.5.1 The SPU-RJ and the Porto Maravilha Project

Reflecting the moment of approximation between the three spheres of the government, result of a convergence in their agendas, and of the mega sporting events, the project formulated by the then mayor of Rio de Janeiro, Eduardo Paes, and his team, was released in June 2009. The main role of the SPU-RJ was to make available the Union's lands to the municipal government, as well as conducting the regularization of their legal and notarial situation. On the SPU-RJ routine of that time, Marina Esteves, superintendent of the state agency in the 2008–2013 period, comments:

> [...] I started to have a team, and then a national recommendation of the central body arrived: 'Look, the president wants it to happen', so I devoted myself to the Porto Maravilha project during the week, not full time, but I dispatched with Eduardo Paes, I was in charge of the control point of the Porto personnel, the CDURP, we participated by telephone, sometimes we participated in person and there was always something that we really needed to do to put the project in motion (Esteves 2017).

Regarding the definition of the areas included in the project, the economic modeling of the CEPACs indicated the number of lands needed to finance the operation. After the mapping, an intense work of CDURP was initiated in conjunction with the SPU-RJ to measure and regularize the lands. With the explicit guidance of the federal government to "unlock" the legal and land situation of the lands destined for CEPACs, the SPU-RJ played a key role in the execution of the project.

In this mobilization scenario, the technical limitations of the SPU-RJ remained evident; however, the context demanded the investment in cooperation with the municipal government. Managing these issues in his office, Esteves reports that the lack of basic software programs prevented engineers from completing their assessments, and they had to request the City Hall the georeferencing and descriptive memorial of the land. This relationship was classified as "accessible," and the dialogue was often informal because of the speed required for the procedures.

---

[18]The last three SPU-RJ superintendents were interviewed in the 2008–2016 period, as well as representatives of the City Hall and of social movements.

In relation to the lands, the entire area created from the landfill in the Port area constitutes a Marine land[19] and, therefore, are federal property—these assets may be linked to municipalities and other federal agencies of the indirect administration, as is the case of some land in the context of Porto Maravilha belonging to *Companhia Docas do Rio de Janeiro* and the Federal Railway Network (RFFSA). The Marine land, margins and what has been added to it, are inalienable and, therefore, third parties can only acquire their eminent domain, which corresponds to 83% of the full domain of the good (the remaining 17% are property of the Union).

The economic valuation of the land needed for the project was under responsibility of the CEF. However, according to Esteves, the consensus was not immediate:

> The initial difficulty were the valuations themselves since we had to evaluate through *Caixa* and it took too much time, but finally […] There was an economic valuation made by the City Hall that we could approve or not, could be an economic valuation made by the SPU or it could be an economic valuation made by *Caixa* because the SPU and *Caixa* had an agreement. Or one could use the valuation made by *Caixa* and use it for the real estate sale. So, at first this was kind of confusing […] Confusing because there was a divergence of techniques. Some divergences were very difficult, because you take a sample that has no similar, then the valuation methodology is what the discussion is really about, and it is a discussion everywhere. *Caixa* evaluated, the SPU thought *Caixa* was undervaluing it and then the discussion began, then *Caixa* said that it was not undervaluing it, sometimes the SPU presented an upward valuation […] Difficult… (Esteves 2017).

When Esteves left his position at the SPU-RJ, Eduardo de Moraes took over as superintendent in October 2013. Moraes also comments that divergent points occurred in relation to the economic valuation made by CEF:

> The lands of the SPU can be evaluated by the SPU, it has legal authorization for that. […] When there are properties owned by entities of the indirect administration or state properties, we can do this if it is in our interest, we are not obliged to do it. The legal obligation is for the entity to conduct a valuation, it is up to it to hire someone to do it and such. Sometimes the SPU helps and makes the valuation, but it is not always. [In the case of *Docas*] It was hiring the CEF and then, I don't remember, there was a problem, a conflict of interest. Because of the CEPACs, then this can give rise to interpret that the guy is undervaluing the land, because the guy wants the land to be sold so that he can then buy his CEPAC. So, there was a discussion there about that. […] In the case of the Gasometer, the first economic valuation created some rumor, we thought we could charge more, but then it went over the top and the City Hall realized it was worth less than that (Moraes 2017).

On the conflict of interest mentioned by Moraes, in 2016 the Public Agency published a newspaper report highlighting that two properties of the Union had been undervalued by the CEF (Belisário 2016). The same CEF, as previously seen, is the fund manager that negotiates the CEPACs and, therefore, it could benefit from the undervaluation of the lands in the purchase to profit later at the sale. In 2010, the time of negotiation of the two lands between the municipal government and the SPU, the latter contested the valuation, resulting in an adjustment, upward, of the values.

---

[19]Marine land comprises a range of 33 m along the coast and the shores of rivers and lagoons that suffer the influence of the tides.

Moraes also reported that some land was traded with the construction of buildings in favor of the Union as a means of payment. However, this arrangement is not positively evaluated:

> In the case of the Gasometer, for example, it had to construct a building as payment. Only then, when I got there, the contract, that had already been signed, established that we were going to build on a land that already had another building, so you had to put down this building, only it was not empty, it still had servers, you had to find a place for the servers that were in that building. […] So it generated a lot more trouble than anything else. But that's it, these construction arrangements [as payment for land] at the beginning seemed cool, but then we saw that cheap was becoming expensive. They were not the best opportunities (Moraes 2017).

Table 5.1 brings an overview of the main lands destined to the city of Rio de Janeiro in the context of Porto Maravilha.

Regarding the actions that would characterize a socio-environmental destination for the lands, Moraes highlights that his conduction to the position of superintendent occurred precisely with the intention of establishing the agenda of the PMCMV Entities in the SPU-RJ. Considered a priority policy since it was inserted in the *Programa de Aceleração do Crescimento* (PAC) [Growth Acceleration Program], it was in the interest of the Ministry of Planning that social interest housing would be effectively implemented. In the port area there was already federal lands mapped to be destined for the PMCMV entities, such as the Quilombo da Gamboa and the Mariana Crioula (both with the destination defined before the start of Porto Maravilha), and only what remained was the completion and signing of the legal processes by SPU. As Moraes points out:

> In Porto Maravilha there was no land that had not been provided for [land for the PMCMV Entities]. What I did was to be able to deliver, because all the contracts moved back and forth with no result. And I managed to regularize all the contracts and sign the contracts with the entities. I think I signed a lot, I think 13 [contracts for the allocation of real estate to the PMCMV Entities]. […] I was already supported by the institutionality provided by the Working Group, so this made things simpler to me because I did not have the work to organize, every month it was all there, and I had only to do my job and deliver it to the desk (Moraes 2017).

The working group cited is the *Grupo de Trabalho Estadual* (GTE) [State Working Group] implemented by the SPU in 2009. In Rio de Janeiro, Sandra Kokudai, of the Bento Rubião Foundation, was one of the GTE participants and reports that only from 2013 the group fulfilled its purpose, following the procedures provided in the constitutive ordinance, performing technical visits across the State and giving the responses demanded by the social movements (Kokudai 2017). This shows that the GTE activities and the dialogue with the social movements and other segments of civil society have their implementation very much associated with the superintendent's initiative and are also consequently affected by the agenda defined by the central body.

Table 5.1 Main lands of the Union negotiated in the context of Porto Maravilha

Land	Owner	Lenght in meters (m^2)	Value of the eminent domain (R$)	Year of sale	Instrument	Form of payment
Formosa Beach	RFFSA	116.125,00	53.108.905,07	2012	Onerous public government leasing	Deposit
Pátio da Marítima	RFFSA	23.809,00	19.380.500,00	2011	Onerous public government leasing	Deposit
Gasometer	SPU	113.209,33	226.300.000,00	2013	Public government leasing under special conditions	Construction work
Usina de Asfalto	SPU	14.603,04	41.000.000,00	2011	Public government leasing under special conditions	N/A*
Barão de Tefé, N° 27 and Venezuela Avenue, N° 154/156	SPU	2.558,41	14.208.770,00	2015	Public government leasing under special conditions	Construction work
Clube dos Portuários	Docas	32.240,00	55.817.500,00	2014	Expropriation upon payment	Deposit
Galpão do Aplauso	Docas	15.021,18	21.810.627,72	2014	Expropriation upon payment	Deposit

Source Own elaboration based on the processes consulted at the SPU-RJ (2017)

* The land called *Usina de Asfalto* was ceded precariously to the municipality of Rio de Janeiro in 1910, and its assignment was officialized by the State Decree N°. 507 of 1975. In 2011, the *Advocacia Geral da União* (AGU) [Federal Attorney General Office] understood that the legal formalization of the leasing should be performed

5.5.2   Institutional Relations and the SPU-RJ Trajectory

Based on the interviews, the SPU guidelines suggest that the trajectory desired by the central body from 2010 focused on Porto Maravilha. Among the mechanisms used to cultivate this trajectory, is the formation of a team within the SPU-RJ office to prioritize the project negotiations, besides the issuance of laws and decrees that would facilitate the transfer of real estate from the Union to the municipality.

In this sense, the performance of the SPU-RJ in the project was based on two main axes: the legal, notary and land regularization of federal public lands and their destination to the City Hall. To enable the state agency to act on these two fronts, certain legal instruments were required, which was provided by the federal government through the promulgation of laws and decrees,[20] and it is in the scope of this legislation that resides the land structure that is vital to the project, especially because its financial engineering was based on the valorization of the urban soil.

In relation to the financial transactions of the lands, the municipal government either requested discounts on the final values claiming that the lands were of public interest or it attempted to negotiate differentiated conditions for payment. On this, it is interesting to verify that in the sale of at least three lands (Gasometer, Venezuela Avenue N° 154/156 and Barão de Tefé N° 27) the payment due by the municipality of Rio de Janeiro was transformed into construction works in favor of the Union. In the contracts of these three lands were foreseen insurance policies equivalent to 10% of the value negotiated in the contract.[21] Another point that draws attention to contracts is the exemption from the payment of the *laudemium* (transfer rate of the eminent domain of the Union's real estate) and the rate of the leasing (annual fee on the eminent domain of the property) based on Law N° 11481 of 2007. Regarding the transactions being beneficial to the SPU, the representative of the SPU-RJ in the 2013–2015 period comments:

> I think some mistakes may have been made, but not from the financial standpoint, but issues regarding the fact of making better arrangements of construction, of smaller or fractional things, or 'I release part of the land and you deliver part of the business', more tangible things. But it had a very heavy political issue at the time there, before my arrival, so many things were done in a hurry because they had to be accomplished and sometimes you would 'intubate' the arrangements that did not generate financial losses, but could obtain better benefits (Moraes 2017).

According to the same interview, in 2013 the performance of the SPU-RJ was strongly directed to social interest housing. As already pointed out, the period is considered one of the most active moments of the GTE in Rio de Janeiro, resulting in the destination of seven properties for the purpose of land regularization and social

---

[20]The main legal prerogatives are: Law N° 11483 of 05/31/2007; Law N° 12348 of 12/15/2010; and the Decree of September 10, 2013.

[21]This type of guarantee-insurance is provided in Law N° 8666/93 to ensure the execution of works in the exact terms agreed with the government, and its value is limited to 10% of the contract value.

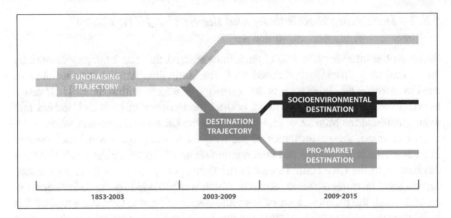

Fig. 5.1 Institutional Trajectories of the SPU-RJ in the 1854–2015 period. *Source* Own elaboration

interest housing,[22] in addition to the publication of nine ordinances that declared areas of the Union as being of interest to the public service for the purposes of land regularization and housing provision, within the scope of the PMCMV entities. This information suggests the cultivation of a trajectory focused on the socio-environmental destination, although it did not interfere directly in the construct works of Porto Maravilha.

As some neo-institutionalist authors theorized and the case study demonstrates, the attempt to change may result from the actions of endogenous actors, when attempting to generate alterations in reproduction mechanisms. Such alterations depend on the cultivation of the mechanisms of a given trajectory, through the mobilization of resources and ideas, to subvert the previous path or at least to reconcile it with the new trajectory. In view of the above, we can map the SPU-RJ trajectories (Fig. 5.1).

Figure 5.1 indicates that the original trajectory remains active in the state government body, taking the form of an action addressed to land demarcation, aliquot definition and rate collection on property use. From 2003 onwards there is a bifurcation, due to the cultivation of a destination trajectory, expressed in the new institutional mission of the SPU. As highlighted by one of its documents it is "[…] in the purposes attributed to the real estate of the Union that resides the main focus of the change of vision in relation to the management of such property" (Secretaria do Patrimonio da União 2010, p. 11). This destination trajectory, as shown by this work, has divided itself into two since 2009: socio-environmental and pro-market. The former is essentially related to the assumptions of urban reform, such as the social function of property, fostering actions of land regularization and social interest housing. In the SPU-RJ, the actions within the PMCMV Entities, as well as the GTE functioning, stand out as associated with this function. The latter, the pro-market, manifested itself

---

[22]The Ordinance (*Portaria*) N° 388/2008 of the SPU declared 21 areas for the National Social Interest Housing Fund (FNHIS). Among these areas, three were in the port area of Rio de Janeiro (Brasil 2008, p. 94).

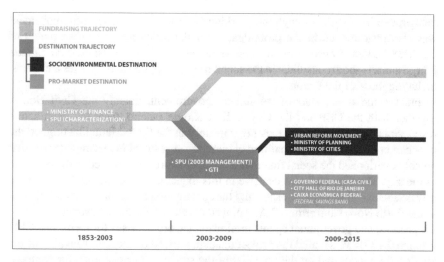

Fig. 5.2 Institutional SPU-RJ trajectories and their actors in the 1854–2015 period. *Source* Own elaboration

specifically in the Porto Maravilha project. Such function was classified in such terms because, according to the narrative, the lands of the Union in the port area served as the basis for the financial engineering of the project and, as a result are subjected to the market. The possible undervaluation of the lands, the type of contracts, the joint action of the SPU-RJ with the municipal power, the approval of specific laws promulgated by the Executive branch of the federal government, are cultivations that favored the private sector.

In order to point out the actors responsible for the cultivation of each trajectory identified here, the following scheme was elaborated the Fig. 5.2.

The main actors in the cultivation of the fundraising trajectory were the Ministry of Finance and the SPU Characterization sector. The former is thus identified by its very function (tax revenue) and its close bond with the SPU. The SPU Characterization sector is described by ex-superintendents as the stronghold of the technicians who conceive the real estate and lands of the Union as its permanent property and therefore must be kept under its possession and serve to gain profit, increasing the tax revenue of the state government body and highlighting the importance of professionals responsible for demarcation and identification of federal assets.

The result of the efforts of the Inter-Ministerial Working Group (GTI) members was essential to inaugurate a new trajectory in the SPU: the destination trajectory. The team that took over the state government body in 2003 sought to put into practice the GTI guidelines through the *Política Nacional de Gestão do Patrimônio da União* (PNGPU) [National Policy for Federal Property Management]. As seen, it was in the implementation of the PNGPU that the socio-environmental destination, also expressed in the new institutional mission, was applied to the properties and lands of the Union. The Ministry of Planning and the Ministry of Cities played an active role at the time (the latter especially by the influence on the participatory management inaugurated in the SPU through the Working Groups). The urban reform movement was also instrumental in the promulgation of the articles of the urban policy of the 1988 Federal Constitution, as well as the approval of the *Estatuto da Cidade* [City Statute]—both consolidated the social function applied to properties and lands, including those of the Union.

Finally, the ruling staff of the federal government, notably the Civil House, together with the CEF and the City Hall of Rio de Janeiro, were the main actors responsible for cultivating the existing trajectory in the Porto Maravilha project: the pro-market destination. The coalition of these actors resulted in serving interests that notably overlooked the social function of the Union property located in the port area of the city. The City Hall was decisive in this trajectory by opting for PPP with the private agents—these agents, especially the construction companies that participated in the Porto Novo Consortium S.A., enjoyed several competitive advantages during the process, and government agents minimized the project risks for them. The CEF, through the FIIPM, secured the project financing, based on the land valuation of the area's public lands and on the potential of the soil as a financial asset for compensating their investors. In addition, it also established a new funding standard, from the FGTS resources, by including urban operations in its investment list. On the other hand, the federal Executive branch, represented by the Civil House, secured the necessary legal instruments, political support and its own assets to execute the project.

In the 2015 Management Report, the SPU pointed out its commitment in recent years to establish its national policy (PNGPU) and pointed out the quantitative and qualitative outcomes of the land regularization initiatives and participation in the My Home My Life Program, as well as other initiatives also within the PAC (Secretaria do Patrimonio da União 2015). This emphasis on destination trajectory is extremely tied to the political demands of the federal government, leaving little room for the implementation of its own guidelines and conferring a reactive character to the management of the Union's assets.

The constant resignifications of the institution and, consequently, of the public policy of the federal property, subject the SPU-RJ to changes in its formal and informal rules, making it an arena where the internal actors (and the external ones) try to combine and cultivate their logic of action with the institutional environment. These resignifications, as a direct result of the political dimension, have ambiguities and contradictions, because they are not necessarily strategic actions. In this sense, the actors assimilate and exploit these incongruences, proper to the institutional systems, in order to transform them into means to reach their preferences and thus influence public policy outcomes which consequently carry these same paradoxical characteristics.

5.6   Conclusion

The new management that took over the SPU in 2003 defined a new institutional mission and a new policy for the management of the federal property, seeking to change the trajectory, originated in 1854, based on earning profit from the use of buildings and lands of the Union. This fundraising function was maintained in the SPU, but the new managers tried to reconcile it with the destination of the federal property for the fulfillment of the socio-environmental function and support to national strategic programs. The insertion of this new trajectory within the state government body signaled an institutional change (albeit limited), characterized as a social process whose dynamics and outcomes directly influence public policies.

As the neo-institutionalist literature showed, the dynamics of increasing returns can also be used to explain the processes of institutional change. Change is not limited to the occurrence of external shocks because the evolution of the institution itself can generate changes in the reproduction mechanisms: internal actors can cultivate the mechanisms of a certain trajectory until they are strong enough to subvert the previous path. In the case of the SPU, the cultivation from 2003 onwards resulted in a trajectory focused on the destination of the federal property. However, in its announcement, the maintenance of the body's fundraising logic was highlighted, occurring what Deeg (2005) called bifurcation. In this case, there is a "limited change" in the institution, since two trajectories coexist and generate increasing returns for the internal actors.

The initiative of the municipal government to revitalize the port area of the city of Rio de Janeiro projected the Union (and consequently the SPU) as a strategic actor for the project. Among the various proposals, its role has undergone changes: in a more recent context, the federal government has ceased to promote housing, including social interest housing, to cultivate a trajectory focused on the land valorization of the Union itself. In the project launched in 2009, it was the SPU-RJ that regularized the federal lands circumscribed in the area and made them available to the municipal government—which subsequently sold them for the same price they had been purchased.

The document analysis and the interviews revealed that the political context, as a variable capable of promoting change, resulted in a cultivation format, the coalition, able to re-signify the destination trajectory and generate a new bifurcation from 2009. The destination ceased to meet the socio-environmental function and incorporated a logic called here pro-market: the public lands served as the basis for the financial engineering of the project, being subject to the financial market. The possible undervaluation due to the economic valuations carried out by the CEF itself (part interested in the purchase of the lands), in addition to the type of contracts, where a great value (about R\$ 240,508,770 million) was linked to the construction of works by the municipal government, favors the use of the public land as a commodity.

In general, the Porto Maravilha project was implemented due to the role of the State: this was the great driver of the process. The performance of CEF, responsible for the monetization of the FGTS and the promotion of real estate development in the region, highlights its approximation with the process of financialization, transforming real estate investment into financial investment. In conjunction with the federal government strategies in the period, which contributed to strengthening private actors and their financial interests, the result was an insulated and complex process, implying little participation, transparency and social control.

The changes in the destination trajectory of the SPU-RJ indicate that despite the intense cultivation to establish it, its meaning has been transformed. The strong centralization and link to the interests of the federal government result in operating logics that, even if they are not completely incompatible, end up in a dispute at certain times, fostering the constant formation of arenas where the internal and external actors attempt to combine and cultivate the logics to which they are affiliated with, thus producing changes in public policy outcomes.

References

Arthur WB (1994) Increasing returns and path dependence in the economy. University of Michigan Press, USA

Belisário A (2016) Engenharia financeira subvalorizou terrenos públicos no Porto Maravilha. Agência Pública, 9 ago, 2016. http://apublica.org/2016/08/engenharia-financeira-subvalorizou-terrenos-publicos-no-porto-maravilha/

Brasil (2008) Portaria nº 388. Diário Oficial da União, 21 out

Caixa Econômica Federal (2017) Minha Casa Minha Vida – Entidades

Crouch C (2005) *Capitalist diversity and change:* recombinant governance and institutional entrepreneurs. Oxford University Press, Oxford

Datafolha (2009) Aprovação a Lula atinge 72%, a maior desde o início de seu governo. Datafolha Instituto de Pesquisa, 21 dez. 2009. http://datafolha.folha.uol.com.br/opiniaopublica/2009/12/1222228-aprovacao-a-lula-atinge-72-a-maior-desde-o-inicio-de-seu-governo.shtml

David AP (2007) Path dependence: a foundational concept for historical social science. Cliometrica 1(2):91–114

Deeg R (2005) Change from within: German and Italian finance in the 1990s. In: Streek W, Thelen K (eds) *Beyond continuity:* institutional change in advanced political economies. Oxford University Press, Oxford

Eduardo Paes é eleito prefeito do Rio de Janeiro (2008) G1, São Paulo, 26 out. 2008. http://g1.globo.com/Eleicoes2008/0,,MUL832458-15693,00-EDUARDO+PAES+E+ELEITO+PREFEITO+DO+RIO+DE+JANEIRO.html

Entenda o negócio (2013) *Porto Maravilha*, 7 Nov 2013. http://www.portomaravilha.com.br/noticiasdetalhe/3981

Esteves M (2017) Interview given to Tuanni Rachel Borba. Rio de Janeiro, March 21

Fioretos O, Falleti T, Sheingate A (2016). Historical institutionalism in political science. In: Fioretos O, Falleti T, Sheingate A (eds) The oxford handbook of historical institutionalism. Oxford University Press, Oxford

Kokudai S (2017) Interview given to Tuanni Rachel Borba. Rio de Janeiro, March 17

Loureiro MR, Abrucio FL, Rosa CA (1998) Radiografia da alta burocracia federal brasileira: o caso do Ministério da Fazenda. Revista do Serviço Público 49(4):46–82, 1998. http://bresserpereira.org.br/Documents/MARE/Terceiros-Papers/98-Loureiroeoutros49(4).pdf

Mahoney J, Thelen K (eds) (2010) Explaining institutional change: ambiguity, agency and power. Cambridge University Press, Cambridge

Ministério das Cidades (2005) Reabilitação de centros urbanos

Moraes E (2017) Interview given to Tuanni Rachel Borba. Rio de Janeiro, January 18

Orren K, Skowronek S (1994) Order and time in institutional study: A brief for the historical approach. In: Farr J, Dryzek JS, Leonard ST (eds) Political science in history: research programs and political traditions. pp 296–317, Cambridge University Press, Cambridge.

Page ES (2006) Path dependence. Q J Polit Sci 1(1):87–115

Pierson P (2000) Increasing returns, path dependence, and the study of politics. Am Polit Sci Rev 94(2):251–267

Reschke A (2010) O Estatuto da Cidade e o papel do patrimônio da União na democratização do acesso à terra e na democratização do estado. Universidade Federal de Minas Gerais, Minas Gerais, Specialization Monograph

Rio de Janeiro (2008) Relatório Final – GT Portuária. Diário Oficial do Município do Rio de Janeiro, 2 janeiro

dos Santos Junior OA, da Silva RH, Sant'ana MC (2011) Introdução. In: dos Santos Junior OA, Montandon DT (eds) Os planos diretores municipais pós-Estatuto da Cidade: balanço crítico e perspectivas. Letra Capital, Rio de Janeiro

Sarue B (2015). Grandes projetos urbanos e a governança de metrópoles: o caso do Porto Maravilha do Rio de Janeiro. Master thesis, Universidade de São Paulo, São Paulo

Sarue B (2016) Os capitais urbanos do Porto Maravilha. Dossiê Capitais do Urbano, Novos Estudos CEBRAP 35(2):79–97, 2016. http://novosestudos.uol.com.br/wp-content/uploads/2017/09/05_BetinaSarue_dossie_105_p78a97.pdf

Secretaria do Patrimonio da União (2010) Balanço de Gestão 2003–2010

Secretaria do Patrimonio da União (2015) Relatório de Gestão do Exercício

Streek W, Thelen K (2005) Beyond continuity: institutional change in advanced political economies. Oxford University Press, Oxford

Thelen K (1999) Historical institutionalism in comparative politics. Annu Rev Polit Sci 2:369–404

Werneck MGS (2016) Porto Maravilha: agentes, coalizões de poder e neoliberalização no Rio de Janeiro. Master thesis. Universidade Federal do Rio de Janeiro, Rio de Janeiro

Chapter 6
Porto Maravilha: The Utopia Crisis After the Global Euphoria

Humberto Meza

Abstract The second phase of the Porto Maravilha project went through an inflection period in 2017, when the consortium Porto Novo—private entity and operator of the venture—paralyzed the maintenance and operation services in the area. Porto Novo demanded the CDURP, a company of Rio Municipality responsible for managing the whole operation, for R$ 6 million (reais) in the court, alleging breach of contract. Despite the services reactivation in November of 2017, this episode uncovered a break in the political coalition that manages the country's largest urban operation and raised questions about the risk distribution. The serious financial crisis in 2015 caused FIIPM illiquidity, the fund responsible for resources. For this reason, the Municipality paid the cleaning and maintenance of the area, dismantling the official speech that the arrangement wouldn't jeopardize public resources. This article intends to explain two elements. On the one hand, we want to test the hypothesis that the coalition rupture has brought to the fore a diversity of actors, in addition to the public–private coalition. The second is that the state acts as an enabler of the process, a sort of "state by contract" that guarantees the financial security of private agents, to the detriment of local resources.

Keywords Porto maravilha · Urban coalitions · Crisis

6.1 Introduction

"The crystal is cracked. I don't know how the relationship will be done from now on, but is not much possible to glue the crystal, no." This brief phrase by Antônio Carlos Barbosa, President of the *Companhia de Desenvolvimento Urbano da Região do Porto do Rio de Janeiro—CDURP* [Rio de Janeiro Port Region Urban Development Company] during the field research for this article summarizes the moment of the

H. Meza (✉)
Observatório das Metrópoles [Observatory of Metropolises], Institute of Urban and Regional Planing, Federal University of Rio de Janeiro, Rio de Janeiro, Brazil
e-mail: meza.humberto@gmail.com

© The Editor(s) (if applicable) and The Author(s), under exclusive license to Springer Nature Switzerland AG 2020
L. C. de Queiroz Ribeiro and F. Bignami (eds.), *The Legacy of Mega Events*, The Latin American Studies Book Series,
https://doi.org/10.1007/978-3-030-55053-0_6

crisis that characterizes the Porto Maravilha operations since 2017, a project in the port region of Rio de Janeiro, the largest Brazilian urban operation.

At that moment, the execution of the second phase of Porto Maravilha works was going through a serious moment when the concessionaire Porto Novo—a private entity and operator of the main infrastructure operation—paralyzed the maintenance and operation services, demanding CDURP in the court (municipality company responsible for the urban operation at whole) for 6 million reais, alleging breach of contract. Although the services were reactivated in November of that year and suspended again in the second half of 2018, this episode uncovered a break in the political coalition that manages the largest urban venture in the country and raised a question around the distribution of risks that characterize the project.

The standstill was a decision by the concessionaire members (composed of three major Brazilian builders, Odebrecht S/A, OAS, and Carioca Engenharia) as a response to the paralyzation of payments for the continuity of the works, when the fund responsible to provide the resources, the *Fundo de Investimento Imobiliáio Porto Maravilha,* FIIPM [Porto Maravilha Real Estate Investment Fund] decreed financial illiquidity affecting the entire operation.

The financial crisis in 2015 (deepened due to the falling in the value of royalties associated with a slowdown in the real estate market in Rio de Janeiro that diminished its power to acquire Certificates of Construction Potential and affected the development according to the original plan), led the FIIPM decreed illiquidity, passing all the obligation to the Municipality (in this case the CDURP), which ended up putting resources to maintain the cleaning and maintenance services of the area.

However, the crisis in the coalition is not only a reflection of a cyclical crisis, but is also part of the arrangement that characterizes this particular Public–Private partnership. To understand the configuration of this impasse, it is necessary to understand the organizational arrangement that governs the urban operation as well as its own financial structure.

The Porto Maravilha project was officially launched in 2009, during the administration of Mayor Eduardo Paes (PMDB), inspired by a set of similar urban operations, such as Puerto Madero in Buenos Aires, the port transformation in Hong Kong, Baltimore, and Barcelona, whose operations were based on the discourse of rescue and revitalization of decaying and abandoned areas in urban centers and port regions, despite the cultural scene and popular that remains in the area.

The first decisive step was the modification of the 1992 Master Plan, through the approval of two municipal laws that allowed, on the one hand, the definition of an Area of Special Urban Interest in the Rio de Janeiro Port Region and in which it would be authorized the Consortium Urban Operation of the Rio de Janeiro Port Region by complementary law 101/09; and on the other, the creation of the Rio de Janeiro Port Region Urban Development Company (CDURP) by municipal law 102/09. CDURP is established as an autonomous company but financially depends on the municipality of Rio de Janeiro (City Hall is the only quotaholder), with the fundamental responsibility of managing the consortium operation.

Some data express the significant importance of this project in terms of population and scope. The project affects an area of 5 million square meters involving seven neighborhoods of downtown Rio de Janeiro (Gamboa, Santo Cristo, Saúde and some sections of Centro, Cajú, Cidade Nova, and São Cristovão) as well as specific intervention programs in *favelas* in the area such as Morro da Providência and smaller operations at Morro da Conceição and Morro do Pinto. This area covers an estimated 32,000 residents, but City Hall estimates a population growth in this region of at least 110,000 inhabitants over the next 15 years, the time for the end of the construction concession granted to the Porto Novo Consortium, accordingly estimated by contract.

With the definition of the area and the constitution of the CDURP, the next step was to define the financial modeling of the operation. According to the report of the actors in the field research, the central point of this model was to attract the real estate market to invest in the revitalization of an area with very low-economic dynamism. This is why the Municipality, through CDURP, decides to issue Additional Construction Potential Certificates (*Certificados de Potencial Adicional de Construção*, CEPACs). CEPACs are titles traded in the real estate market that allows the real estate sector that acquires them, the ability to build projects outside of the local building regulations, but conditioned to the requalification of a specific area (Werneck 2016). In other words, the availability of CEPACs would make the revitalization of the port region completely attractive to the real estate market, driven by the logic of commercialization of public land in that particular area.

With the issuance of CEPACs and the availability of public land, CDURP has the resources to start administrative operations, but the issue of the availability of sufficient land for the whole project emerges as an unavoidable problem (Pereira dos Santos 2016).[1] According to the speech of different actors in the field, more than 30% of the land in this area belonged to the Federation, so it was necessary a political pact between the three levels of the Federation (Municipal, State and Federal) in order to guarantee that those lands could be available to the real estate market.

To start the operations, the City sold to the Federal Government the first package of 400 thousand CEPACs for a value of 3.5 billion reais. The money was used to capitalize CDURP, which used all these resources into the creation of an investment fund, the Real Estate Investment Fund of the Port Region [*Fundo de Investimento Imobiliário da Região do Porto* (FIIRP)]. Then, still in 2010, FIIRP made a public offer to sell CEPACs to the market for a total of 7.6 billion reais. The Time of Service Guarantee Fund [Fundo de Garantia por Tempo de Serviço, FGTS],[2] fund managed by the public bank CAIXA Econômica Federal, appears as interested buyer of the total of CEPACs in a single lot.

---

[1] To understand the debate about the availability of federal land transferred to the Municipality for the Porto Maravilha operation, we recommend Tuanni Borba's chapter "Institutional Analysis of the Secretariat of Federal Property (SPU)" in this book.

[2] FGTS is a fund that belongs to Brazilian workers as guarantees of unfair dismissal or any cause of unemployment. It is a fund guaranteed by employers and corresponds to 8% of the salary of each worker. The FGTS is administered by the Federal Government through Banco CAIXA Econômica Federal. CAIXA has the faculties of managing and investing in real estate funds, ensuring the provision of funds for certain urban programs and policies, such as the *Minha Casa, Minha Vida* Program, for example.

FIIRP agrees to sell the total CEPACs to the FGTS in a single lot, which is expected to pay the venture's 7.6 billion reais plus CDURP's administrative expenses of 400 million, thus totaling the operation in 8 billion reais over 15 years of the project. To manage this operation, FGTS creates the Porto Maravilha Real Estate Investment Fund (FIIPM), which is responsible for direct negotiation of the set of CEPACs with the real estate market.

From here, the scheme of the resources transferring for the maintenance of the works is relatively simple to understand. FIIRP signs the operation agreement with the Porto Novo Consortium, which assumes responsibility for maintaining the area, building the main infrastructure, and other urban devices oriented by CDURP.[3] For each operation, the Consortium issues a service invoice to FIIRP. FIIRP sends the invoice service to FIIPM who must deliver the money to pay for the selected transaction. In other words, CDURP through FIIRP coordinates the entire operation, but is not responsible for the resources.

The scheme worked smoothly until May 2016. Three months before the Olympic Games, FIIPM informs CDURP (FIIRP Administrator) that it had no liquidity to maintain cash flow. According to Barbosa:

> Then the problems began. In the transition of (municipality) government (starting 2017) it had a commitment that City Hall would put the money in the CDURP in order to CDURP negotiate with the FIIPM and buy CEPACs again. In the past, when the Municipality bought CEPACs directly with the FIIPM, it generated money to the FIIPM and then, we paid to Porto Novo (…) the government made several transfers, but didn't guarantee budget in the Municipality for it. We were missing 219 million reais and we couldn't pay for it, so they, Porto Novo suspended the services.

> (Interview with Antônio Barbosa, President of CDURP).

In consequence, not only the Consortium interrupted all the project operations, but also CDURP had to use the resources and companies linked to Rio's City Hall to guarantee maintenance and cleaning operations, thus dismantling the official discourse that the project's organizational arrangement would not commit public resources to guarantee the operation. At the same time, the crisis revealed that the public–private coalition that manages the venture is also shaped by tensions, discarding the assumption that partnerships linked to urban operations remain unchanged throughout the whole process.

Starting from this episode, this chapter analyzes the configuration of the public–private coalition through two central arguments. First, we argue here that urban operations guided by the real estate commercialization logic go through cyclical turbulences that weaken the coalition and reveal an inequity in the distribution of venture risks. This impacts, second, on the way the state organizes itself to deal

---

[3]During the first phase of operation, from 2010 to 2016, this model allowed the development of initial works such as the construction of two road tunnels, the overthrow of Perimetral Avenue, the construction of two museums (Rio Art Museum and Museum of Tomorrow), 70 km of reurbanized roads and 650,000 m^2 of sidewalks, among other related works. The second phase would be characterized by the construction of large real estate developments but it is the phase that faces the most profound stagnation as we will analyze in this chapter.

with these conjunctural uncertainties. Beyond a state that guarantees the operation, what we observe from this case is that not only public agents abdicating their role of control over private actors, but also they assume the function of enabling the process, enacting a kind of "state by contract" (Pereira dos Santos 2016) that guarantees the financial security of private agents, to the detriment of even local resources.

It is not here an analysis of the financial arrangement for this kind of urban operation (Mosciaro et al. s/n), but what are the changes and reorganizations in the public–private coalition that manages these type of projects at critical junctures (Pierson 2000) and, fundamentally, what is the state configuration to deal with the risks present in those situations.

To conduct this debate, the chapter is divided into three sections. In the first, we want to analyze the strategies, perception, and movements of the members of the coalition (FIIRP/CDURP, FIIPM/FGTS and Porto Novo Consortium) from the crisis scenario that marks the stagnation of the operations of the second phase of Porto Maravilha.

In the second section, from this case we will dialogue with the literature of political coalitions, adopting the neoinstitutionalism approach that analyzes the role of the agents trying to argue that such strategies also respond to a trajectory of the relations between the state and the private sector inscribed in the democratic process Brazilian. In other words, the crises in the coalition repeat a dynamic already learned in public–private relations in Brazil, but with an innovation inscribed in the current moment of deepening the neoliberalization process (Ribeiro 2017).

Finally, in the last section, we will try to draw conclusions about the contributions offered by the public–private coalition that manages Porto Maravilha and to understand how much it dialogues with the debate about coalitions that govern the urban operations PPPs.

6.2 Configuration of a Coalition in Crisis

The revitalization history of Rio de Janeiro's port area is, among other things, a history of the state's challenges in gaining the interest of the real estate market to bet on urban developments in the port area. Several researches (Giannella 2013, 2019) show that at least seven revitalization proposals (most of them little or not materialized in part also due to the lack of alignment of the three levels of government) in the port region have emerged since the early 1980s with massive removal strategies in the surrounding neighborhoods thus stimulating resistance and popular mobilizations.

The initiative begins to become viable as early as 2004, under the administration of Mayor Cesar Maia. During the end of his term in 2007, Maia made a public call for a financial feasibility study of the revitalization project in the port region for which a consortium of construction companies formed by Odebrecht S/A, Carioca Engenharia, OAS, and Andrade Gutierrez, practically the same actors that in 2009 constitute the Porto Novo Consortium, which is responsible for the revitalization and maintenance works of Porto Maravilha (except for the Andrade Gutierrez company).

This feasibility study resulted in the creation of a financial modeling proposal delivered to the, at that moment candidate Eduardo Paes (PMDB), which assumes it as a campaign promise.

The Paes election as Mayor of Rio was extremely important for the government alignment (Pereira dos Santos 2016). He was elected within a colligation led by the PMDB. One year before, the PMDB was also elected for the Rio State Government in name of the Governor Sergio Cabral and, at the same time, the PMDB was also an important political allied of the ex-President Lula da Silva at the federal level. This political alignment was so important for the Porto Maravilha venture, reinforced by the selection of Rio for the final game of the World Cup in 2014 and the Olympic Games in 2016.

Although Municipal and CDURP officials argue that the revitalization works in the port region are not part of the city's preparations for the 2016 Olympic Games, the selection of Rio as an Olympic city has spurred the acceleration of the venture, particularly after the International Olympic Committee (IOC) has approved the operation of several Olympic-related operating areas in that region, such as Referees Village, Media Village, Olympic Boulevard, Hotels, etc., thus making the region completely attractive to the real estate market (Giannella 2019).

As researcher and doctor in financial law Alvaro Pereira dos Santos analyzed during our field research, it was a process in which the global spotlight on Rio de Janeiro converge with an alteration of the urban expansion process that "returns" to the central area with proposals for projects oriented under a financialization sense:

> Undoubtedly, the Olympic Games stimulated all this urban operation, but there's an earlier issue. The Olympic Games would not have happened in Rio if it hadn't been a growth process that, in some way, became Rio an interesting place to host the Olympic Games. So I wouldn't see the Olympics as the main cause for the escalation of this real estate valuation, from which Porto Maravilha starts. Of course there's a worldwide trend on the return of the real-estate market to Downtown and it isn't necessarily a replacement of one expansion front by another, but a growth of a city scale as a whole. Although the Eduardo Paes speech always emphasized this operation as a strategic option for redirecting the urban expansion to Downtown, this never implied the abandonment of the Barra da Tijuca vector. Since the beginning of the growth cycle in Brazil in 2007, Barra is chosen as a priority area for the Olympic Games, the expansion of the Subway and the BRT System. At the same time that all those urban operations are running in Barra, they repeat the discourse that expansion is problematic and we must to return to the Center Area. It's a completely ambiguous speech. They don't need to question an expansion that, in any case, continues to happen.

(Interview with scholar Alvaro Pereira dos Santos, researcher)

In terms of the confirmation of the public–private coalition that manages the venture, we observe that the actors also reproduce a previously existing coalition model. Not only the Porto Novo Consortium actors are the same members of the previous consortium who delivered to Eduardo Paes—during his campaign as candidate to Mayor—a financial modeling proposal for the revitalization of the port region that ended up being similar to the Porto Maravilha modeling defined in 2009–2010 as well as the same private agents that participate in a series of PPPs that accentuate the urban expansion process of the city in the centralized vectors of South Zone and Barra da Tijuca.

During our field research, we had access to the urban operation projects in which the construction company Odebrecht S/A participated through several contracting models, from direct bidding through different PPPs to the Urban Consortium Operation that specifically characterizes the Porto Maravilha project. Among these major works carried out by Odebrecht, whether in direct bidding or via PPPs, we can list the renovation of Maracanã Stadium, the construction of a new terminal at Galeão International Airport, the construction of the Olympic Park, the construction of the transportation system BRT and its main road (Transolimpica, Transoeste, and Transbrasil, the latter still under construction) as well as participating in the PPP for the VLT (Light Rail System) that runs through part of the main stretches of Porto Maravilha. In this way, Odebrecht becomes a key player in urban operations, with a clear ability to influence a range of decisions related to the creation of urban space that involves private actors.

The first serious crisis affecting the coalition, marked by the stoppage of services in the maintenance of the Porto Maravilha works and accentuated by the Porto Novo Consortium's legal demand against CDURP, was overcome in November 2017 but distrust remained throughout the period. Since July 2018, when FIIPM was to confirm the payment of the sixth service order, services stopped again after FIIRP confirmed to the consortium that FIIPM has again declared illiquidity. The consortium's stoppage of operations and, therefore, the suspension of the contract have continued since then and until the writing of this article in the first half of 2019 had not been resumed, according to the latest CDURP operations report issued in April 2019.[4]

We spoke with the Odebrecht's Director of Engineering a few weeks before FIIPM's confirmation of illiquidity in the second half of 2018, and from his statement, we can infer that the contractual crisis triggered a present tension in the relationship between the funds (FIIRP and FIIPM), which impacts to some extent on the stability of the coalition as a whole.

> Caixa Econômica Federal realized in this (the venture of Porto Maravilha) an opportunity and they acquired everything (all the CEPACs). For a bank that will live one hundred, two hundred, three hundred years to "put it all in", putting these securities titles in portfolio and monitoring its evolution, it is an investment strategy. But in 2017 we had a very big problem! There came a time when we knocked on the Municipality door and said, "Chief, I cannot do. I don't have money in my company. My company provides service. I don't sell CEPAC. We don't do real-estate development, I do operation, maintenance of the place and build what you give me to build, according to the master plan that exists within the operation. If I don't have money, I stop building. You don't have the money to keep? I stop to keep". And this is a reality. I'm just a service provider.
>
> (Carlos Hermanny Filho, Director of Engineering, Sustainability and Innovation. Odebrecht)

This strategy of "keeping all securities titles in the wallet" that Hermanny comments on is about the decision that CAIXA/FIIPM made to hold the set of CEPACs and not put them on the market. During the beginning of the second phase of the works in Porto Maravilha, FIIPM decides not to sell CEPACs and, instead, to use them as resources to associate with future real estate developments. However, the

---

[4] All the reports by trimester are available directly from the CDURP website at this link: https://www.portomaravilha.com.br/trimestrais/relatorios/.

housing market crisis that began most clearly in late 2016 after the Olympic Games impacted on the lack of developments in the region, which in practice translates into CAIXA's loss of business opportunities. By CDURP's view, through its own president, this is the reason for FIIPM's illiquidity that affected the whole operation, thus revealing a tension present in the interaction between the two public funds that manage the operation.

> CAIXA says "I don't want to sell, but I give you, investor, the land and CEPACs. I want to participate in your business". So, when the crisis arrived, they faced problems. They didn't sell the lands, they didn't sell the CEPACs, and the operation hasn't have liquidity. The illiquidity is consequence from irresponsibility or a wrong investment policy and a wrong administrator.
>
> (Interview with Antônio Carlos Barbosa, president of CDURP)

If the crisis operates as a variable not controlled by the actors, it also reveals holes in the coalition's arrangement. The financial model and the coalition were designed in the middle of a scenario in which Brazil lived in a climate of political normality and permanent growth. If the scenario were normalized, perhaps the contract equation would have been achieved, but the sharp political crisis in 2015 that preceded impeachment to President Dilma Rousseff, associated with the fiscal crisis in the state of Rio de Janeiro showed that the model breaks down.

The continuity of the process was not completely made possible by the sale of CEPACs, contradicting the initial sense that financial modeling would account for the sustainability of the operation (Marques 2008; Werneck 2016), while revealing in practice that there is uneven distribution of the risks.

If we analyze the coalition arrangement, we can observe the existence of four agents. In the end, the real estate developer that makes each specific project, then we find the PPP concessionaire (in this case, the Porto Novo Consortium) that makes the infrastructure as a whole, the semipublic funds (FIIPM and FIIRP) that are investing and the agents' public services (CAIXA, City Hall and the Rio State).

In practice, the real estate developer allocates resources gradually and according to the market dynamics. It buys CEPACs, land, makes a specific venture and sells, all at very low risk. The PPP concessionaire, by its side even eventually operates without any loss, suffering only the immediate nonfulfillment of the contract, as it has been doing since 2017. Thus, it is doing the works and receiving according to the schedule, making profit as far as it possible, but without receiving money does nothing else. FGTS, the FIIRP's administrator has a risk of actual damage inherent in the contract, since it is assumed that it could use the resources of Brazilian workers to invest in social housing policies, leaving the possibility that FGTS may even, depend on the crisis, give up on the operation.

In the other point of this chain is the City Hall, which since the first moment of the coalition crisis, it continues to transfer public assets to ensure the maintenance of the operation (assuming expenses and risks in this gear as a whole). From the beginning of the operation, the idea was established that the private actor would be the banker for the entire operation (Werneck 2016; Mosciaro et al. s/n), but even if this had in fact been confirmed in practice, this is not a model that determines the absence of the state. In practice, state entities act as a security predictor, a coordinator of the whole process, allowing the market to see the operation as attractive and necessary to invest, an analysis that is once again present in the analysis developed by Doctor Alvaro Pereira dos Santos during our field research:

> We received the speech that the state would be only a facilitator and the entire operation is feasible by the market. There is a whole discursive ambiguity around the state as facilitator, but what we see here and although all the financial modernity about the CEPAC model, etc., is an old ways of repeating itself with a new outfit. We see the public fund funding, a series of corruption scandals. It creates a whole fiction on the capital market, but in fact you don't have this market of CEPACs developed. What is new, then? It has a whole financial rationality imposing itself on the public power.
>
> (Interview with scholar Alvaro Pereira dos Santos, researcher)

In short, what this arrangement of the coalition and its practice in the current context of crisis can inform us about the debate around the coalitions that define and govern the urban space? How does this debate explain the role of agents and coalitions? This is an analytical operation that we intend to perform in the next section.

6.3 Coalitions and Agents in the Production of Urban Space

While theoretical approaches on the urban issue have shifted focus to emphasize the role of agents in the production of cities, the whole analytical agenda have shown a clear liability of formal institutions as the only way to explain decision-making about urban policies. Indeed, the interdisciplinary field of urban studies has benefited greatly from the concept of "growth machines" formulated by Moloch (1976) and enriched by a fertile research agenda around coalitions (Logan and Molotch 1987) evidenced a lasting relationship between real estate agents and local government institutions for the production of urban policies.

More than the existence of the coalition per se, what this research agenda has shown us that the state's inability to produce the resources needed to implement urban policies pushes cities to become growth spaces stimulated by the aggressiveness of real estate actors (Mota and Fedozzi 2019) to respond to the formulated development demands.

From the late 1990s, the theoretical approach to Urban Regimes (Savitch and Kantor 1997) of Anglo-Saxon genesis and subsidiary of coalition debates strengthened the understanding of public–private interactions by broadening the focus to capture the dimensions of democracy at the local level. The analysis of the forms of distribution of power in the city environment (or in the municipality) stimulated to observe that government decision-making in the management of cities is not "isolated" from the structure itself but takes on the formations of struggle, fields, agenda, of regulations, etc., with several participants, forms of pressure, and rational choices.

Such an approach demanded a necessary theoretical fit (as it needs to explain the formulation of power sharing) and methodological (as it must use its own research tools) adjustment between a reading of the Political Regime—usually, nationally or internationally—in tune, reading the urban processes and agenda, that is, the forms of government decision that affect the production, supply, service provision, among others, of equipment and services whose logic is given in the city, by the city, and for the city.

Thus, both agents and institutional configurations are required as the object of fundamental analysis to explain the urban production process. This constatation would make the neoinstitutionalist approach more attractive for understanding the decision-making process of the urban issue, as Mota and Fedozzi (2019) argued by reading the Lowndes (2009) contributions:

> According to neoinstitutionalism, institutions constrain behavior and are transformed by the agency. Thus, in the analysis of political processes, both informal conventions and coalitions as well as the formal rules and structures that shape political behavior deserve considerable attention, as well as the manner in which political institutions embody values and power relations (2019: p. 7).

The neoinstitutionalist focus, in turn, compels us to understand that agents operate in the institutional field-making calculations aimed at their interests,[5] but those calculations are not elaborated in a vacuum. They operate constrained by the urban context and depend on interactions with other actors and political phenomena. The combination of them all stimulates institutional change that could only be consolidated over time by self-reinforcing mechanisms, thus forming a new institutional pattern (Hall and Taylor 2003: p. 199).

In the case of the Porto Maravilha urban development, analyzed here, it is clear that the agents bet on the public–private coalition for the existing previous relations and the mutual benefits of this relationship, since the Porto Novo Consortium actors were already transiting in the most-varied urban enterprises. The size of the interest of these agents operates, at the same time, as a stimulus to the political dynamic, since the intergovernmental alignments (the three levels of the Brazilian state completely aligned around the operation) and the calculations in the election of Mayor Eduardo Paes allowed some way that the venture would materialize.

However, both the agents' calculations and their interests operate in a clear scenario of uncertainty. In response to the criticism of Hall and Taylor (2003), about the causal mechanism model formulated in the path dependence concept (PIERSON, 2000) of the neoinstitutionalist debate, Pierson and Scokpol (2008) argue that the institutional trajectory is also shaped by the unexpected results of the agent's action.

The unpredictability of the institutional course appears more frequently than we realize, to the point of reaching a regular status in the process of institutional formation. According to Pierson and Scokpol (2008) even if actors act in favor of a specific institutional change, it is clear that they operate in a scenario of high uncertainty and therefore it can be foreseeable that they will make "mistakes" (2008: p. 21).

---

[5]This understanding has been highly consecrated in Political Science from the various meanings of rational choice theory that focuses on actors. This approach came from the economic understandings of political phenomena, thanks to the contributions of Antonhy Downs, Mancur Olson and George Stigler.

Thus, the resulting trajectory of institutional change may have unplanned and there-fore highly contingent elements, implying yet another methodological challenge for understanding the empirical field (Bennet and Elman 2006: p. 252).

Thus, some phenomena of Brazil's after 2013 political–institutional context, such as Car Wash Operation [*Operação Lava Jato*], which delegitimized the Brazilian party system and undermined the investments of large contractors (particularly Odebrecht), operated as factors of uncertainty, generating the venture crisis and ruptures in the coalition that manages the operations. Thus, both agents with their interests and the calculations of their actions, such as the institutional framework and its trajectory, are necessary keys to understand the conditions and results of the coalitions that produce the urban space.

6.4   Conclusion

It is practically consensual that after 10 years of its launch, the Porto Maravilha project is in deep stagnation (Werneck et al. 2018). Along with the work stoppage due to the breach of the contract between CDURP and the Porto Novo Consortium as well as the tensions between the semipublic funds (FIIPM and FIIRP), the CEPACs market does not seem to be viable as sufficient fuel for the breath of operation's constructions.

A quick reading of the second CDURP report of 2019 reveals a very slow dynamic in the sale of CEPACs, with only 8.93% of titles consumed.[6] By December 2018, more than half of the high-level buildings built were unoccupied, while only 50% of the planned intervention works were completed. According to the findings in the field research, to overcome the emptiness of occupations, there was a prediction that financial sector actors (both CAIXA and a private banks) would be planning to move to the region and occupy some built buildings, but until the writing of this article, no movements have been confirmed at all.

There is no forecast for the PPP contract to be resumed unless FIIPM reactivates the transfer of the 147 million reais that allow FIIRP to resume infrastructure and service operations with the Porto Novo Consortium. At the same time, the Rio de Janeiro City Hall continues to have expenditures already estimated at 342 million reais to solve the lack of private sector services in the region.[7]

This stagnation scenario is combined with a series of abandoned urban develop-ments, from the Olympic Park arenas to the road networks that were designed, in principle to prepare the city for global events. The panorama reveals not only the existence of tensions in the coalitions that create the urban space (consequence of the crisis in the interaction between the agents' interests and the institutional trajectory),

---

[6]Report available at: https://www.portomaravilha.com.br/relatorios_trimestrais/.

[7]https://oglobo.globo.com/rio/apenas-50-das-intervencoes-em-infraestrutura-foram-concluidas-no-porto-maravilha-23301634.

but also the difficulty of the agents of the state and local power in guaranteeing the development and dynamism of the largest *Brazilian showcase* in post mega event times.

References

Bennet A, Elman C (2006) Complex causal relations and case study methods: the example of path dependence. In: Advance access publication

Giannella L (2013) A produção histórica do espaço portuário da cidade do Rio de Janeiro e o projeto Porto Maravilha, Espaço e Economia [Online], 3 |2013, posto online no dia 19 dezembro

Giannella L (2019) Financeirização e Produção do Espaço Urbano no Porto Maravilha, Rio de Janeiro. Neoliberalismo às avessas? Paper apresentado no XVIII Enanpur, 2019. Maio

Hall P, Taylor RC (2003) As três versões do neo-institucionalismo. In: Revista Lua Nova, No. 58. São Paulo

Logan J, Molotch H (1987) Urban fortunes: the political economy of place. University of California Press, Berkeley

Lowndes V (2009) New institutionalism and urban politics. In: Davies J, Imbroscio D (ed) Theories of urban politics, 2nd edn. Sage, Los Angeles/london/new Delhi/Singapore/Washington DC, pp 91–105

Marques CN (2008) Para onde vai a zona portuária do Rio de Janeiro?: singularidades do lugar e diversidades de propostas. 2008. 161 f. Dissertação (Mestrado em Geografia Humana)–Instituto de Geografia, Universidade Estadual do Rio de Janeiro, Rio de Janeiro

Moloch H (1976) The city as a growth machine: toward a political economy of place. Am J Sociol 82(2):309–332

Mota VM, Fedozzi L (2019) Atores Urbanos e Arranjos Institucionais nas políticas de "revitaliza-ções" portuárias do Brasil contemporâneo: Notas de uma pesquisa em curso. Paper apresentado no XVIII Enanpur, 2019. Maio

Mosciaro M, Pereira dos Santos AL, Aalbers M (s/n) The finacialization of urban development: speculation and public land in Porto Maravilha, Rio de Janeiro. In: Chu CL, He S (eds) The speculative city: emergin forms and norms of the built environment. Toronto University Press, Toronto

O Globo. https://oglobo.globo.com/rio/apenas-50-das-intervencoes-em-infraestrutura-foram-con cluidas-no-porto-maravilha-23301634

Pierson P (2000) Increasing returns, path dependence and the study of politics. Am Polit Sci Rev 94(2)

Pereira dos Santos A (2016) Intervenções em centros urbanos e conflitos distributivos: Modelos regulatórios, circuitos de valorização e estratégias discursivas. Tese de Doutorado em Direito. Universidade de São Paulo, USP, São Paulo

Pierson P, Skocpol T (2008) El Institucionalismo Histórico en la Ciencia Política Contemporánea. Revista Uruguaya de Ciencia Política. Revista Uruguaya de Ciencia Política 17(1):07–38

Porto Maravilha. https://www.portomaravilha.com.br/relatorios_trimestrais/

Riberio LCQ (ed) (2017) Rio de Janeiro: Transformações na ordem urbana. Letra Capital, Rio de Janeiro

Savitch HV, Kantor P, Hadock SV (1997) The political economy of urban regimes: a comparative perspective. Urban Affairs Rev 32:348–377

Werneck M (2016) Porto Maravilha: agentes, coalizões de poder e neoliberalização no Rio de Janeiro. Dissertação apresentada ao curso de Mestrado do Programa de PósGraduação em Planejamento Urbano e Regional da Universidade Federal do Rio de Janeiro – UFRJ

Werneck M, Novaes P, dos Santos Junior OA (2018) A estagnação da dinâmica imobiliária e a crise da operação urbana do Porto Maravilha. Informe Crítico. Observatório das Metrópoles

Chapter 7
Informality and Invisibility of the Slum Tenements in the Port Area of Rio de Janeiro

Orlando Alves dos Santos Junior, Larissa Lacerda, Mariana Werneck, and Bruna Ribeiro

Abstract The article aims to discuss the first results of the survey developed in the port area of Rio de Janeiro around the existing slum tenements (called *cortiços*, which are several buildings with shared bedrooms where many lower class families live together) in the region, their housing conditions and the profile of their population. Given the situation of informality and lack of information about this type of housing in the diagnosis presented by the municipality in the context of the elaboration of the Social Interest Housing Plan in the second semester of 2015, we sought to fill this gap in a work carried out street by street, identifying the properties that operate as slum tenements in the area of Porto Maravilha Urban Operation. Thus, we sought to deconstruct the current perception that stigmatizes these spaces—and their inhabitants—as precarious and marginal, showing that the slum tenements are marked by a great heterogeneity of housing conditions and social groups, unified in their demand to live in the central area of the city (Research conducted by the Observatório das Metrópoles in partnership with *Central de Movimentos Populares* (CMP) [Central of Popular Movements]. Funding: Ford Foundation.).

Keywords Slum tenements · Informality · Central areas · Housing

O. A. dos Santos Junior (✉)
Institute of Urban and Regional Planing, Federal University of Rio de Janeiro, Rio de Janeiro, Brazil
e-mail: orlando.santosjr@gmail.com

L. Lacerda · M. Werneck · B. Ribeiro
Observatório das Metrópolis, Institute of Urban and Regional Planing, Federal University of Rio de Janeiro, Rio de Janeiro, Brazil
e-mail: larissa.gdynia@gmail.com

M. Werneck
e-mail: marianagsw88@gmail.com

B. Ribeiro
e-mail: cribeirobruna@gmail.com

© The Editor(s) (if applicable) and The Author(s), under exclusive license to Springer Nature Switzerland AG 2020
L. C. de Queiroz Ribeiro and F. Bignami (eds.), *The Legacy of Mega Events*, The Latin American Studies Book Series,
https://doi.org/10.1007/978-3-030-55053-0_7

7.1   Introduction[1]

Since 2009 is underway in the city of Rio de Janeiro, the renovation project of the port area implemented through the *Operação Urbana Consorciada Porto Maravilha* [Porto Maravilha Urban Partnership Operation], managed by *Companhia de Desenvolvimento Urbano da Região do Porto do Rio de Janeiro—CDURP* [Rio de Janeiro Port Region Urban Development Company]. The Urban Operation involves works and services in the 5 million square meters of the *Área de Especial Interesse Urbanístico—AEIU* [Special Urban interest Area] of Rio port area amounting to R$ 8 billion *reais*, implemented through a Public–Private Partnership (PPP) won by *Consórcio Porto Novo* [Porto Novo Consortium] (integrated by the companies *Odebrecht Infraestrutura, OAS* and *Carioca Christiani Nielsen Engenharia*).

The analysis of the interventions foreseen in the Urban Operation reveals the lack of investments in social interest housing, aiming at the permanence of the current dwellers and the expansion of housing addressed to the popular classes. In other words, there are no public resources being invested in housing although an increase in population and in demographic density is expected to occur in the region. According to municipal government calculations, the population would increase from the current 32,000 to 100,000 inhabitants until 2020, living in the neighborhoods of Santo Cristo, Gamboa, Saúde, and in parts of the central area of the city.

The lack of policies and investments in social interest housing led several popular organizations to press the municipal government and CDURP to elaborate a social interest housing plan for the port area, which occurred in 2015 through public hearings and a municipal conference held in August of that year with the goal of discussing and approving this plan.[2]

It is worth noting that the diagnosis set up to serve as input for the elaboration of the said plan does not mention anything about the existence of slum tenements in the port area,[3] although this type of housing in the central area of the city was of general knowledge. Similarly, it is surprising that there is no official information provided by the public bodies about slum tenements, considering the number of indications that shows that such type of housing is expressive and has been spread downtown. The invisibility of the slum tenements in the official diagnoses also prevented discussions on proposals focusing public policies addressed to the slum tenements and their dwellers since their existence was not recognized.

This scenario is even more alarming if we consider the removals conducted by the City Hall during the construction works of Porto Maravilha. According to data

---

[1] This article was originally published in the Journal *the Social in Question*, year XXI, N. 42, Volume 1, September-December (2018), Rio de Janeiro: PUC-Rio, pp. 83–118.

[2] To see the full Social Interest Housing Plan, see http://138.97.105.70/conteudo/habilitacaoSocial.zip. Accessed: October 2016.

[3] Cf.   http://www.portomaravilha.com.br/conteudo/outros/Diagnostico_PHIS%20Porto%20rev%20abr2016.pdf. Accessed: October 2016.

from the World Cup and Olympics Popular Committee of Rio de Janeiro,[4] from 2009 to 2015, at least 535 families were removed from the port area. These removals especially affected occupations organized by housing movements in buildings that were not fulfilling their social function, many of them abandoned decades ago. Such scenario depicts the unwillingness of the city to guarantee the permanence of low-income families in the area, since many of these buildings remain empty. Moreover, the difficulties faced in the access to public information lead to the belief that these figures can be even higher and, according to reports from some dwellers in the area, properties that functioned as slum tenements may have been the target of this removal policy.

In this context, the objective of this article[5] is to summarize the results of the research on the slum tenements in the port area[6] conducted from a field survey, street by street, aiming to identify the slum tenements and the profile of their dwellers. The results confirm the presence and expressiveness of this type of housing in the central area of the city. At the same time, from a historical perspective and critical theory, we sought to discuss the importance of the slum tenements as a way of access to centrality, the reasons for their reproduction as an informal housing over time, even without their regulation by the public power, and the meanings their dwellers attributed to them.

The survey was carried out in the neighborhoods of Santo Cristo, Gamboa, Saúde, and part of the central area of the city included in the Porto Maravilha Urban Operation.

Despite the lack of a precise and objective definition of what a slum tenement is, in general, the existing definitions consider slum tenements "[...] real estate that has as its main characteristic precarious housing conditions resulting in subhuman living and housing conditions..." (Saule Júnior et al. 2007, p. 370). Among the conditions of precariousness, the following situations stand out: (i) physical safety and subdivision in several rented or subleased rooms; (ii) overcrowded rooms, resulting from the number of inhabitants, that is, disproportionate and incompatible with the property size; (iii) rooms with no window or ventilation; (iv) rooms with multiplicity of uses; (v) insufficient number of sanitary facilities revealing precarious hygiene conditions;

---

[4]Available at: http://www.childrenwin.org/wp-content/uploads/2015/12/Dossie-comit%C3%AA-Rio2015_low.pdf. Accessed: November 21, 2016.

[5]The first version of this article was presented at the XVIII National Meeting of the Post-Graduation Association in Urban and Regional Planning, held in 2017 (Santos Junior et al. 2017).

[6]To perform the survey, a fieldwork was carried out, street by street, in which we sought to identify the existence of rooms to rent, asking questions in newsstands, bars, and the slum tenement dwellers. For the collection and organization of the information, three forms were elaborated: (i) the first, filled out by the field agents, gathered information for the identification of the slum tenements, even when it was not possible to interview any dweller, administrator, or owner; (ii) the second was a questionnaire with questions addressed to the owners or administrators, when it was possible to interview them, focusing on information related to the functioning of the establishment; and (iii) the third was a questionnaire with questions addressed to the slum tenement dwellers, interviewed randomly when possible, aiming to establish a social profile of this social group and its living conditions.

(vi) lack of safety of electrical installations, generating fire hazards; (vii) overload of water and electric energy consumption due to overcrowding (Saule Júnior et al. 2007, pp. 370–371).

Although the above definition focuses on the physical conditions of housing, Saule Júnior et al. (Saule Júnior et al. 2007, p. 371) emphasized that "[…] the situation in the slum tenements is not only irregular and precarious in relation to the conditions of habitability. The irregularity and precariousness also exist in the informality of the legal relations between the slum tenements dwellers and the owners of these urban properties." In general, "[…] the slum tenements dwellers, while subtenants, remain in a precarious renting legal relationship."

Given this, we decided to conduct a more restrictive survey of the properties considered slum tenements. Our analysis focuses on the properties that have rented rooms, cohabitated by more than one family, and it excludes forms of collective housing such as occupations which, despite the possibility of presenting physical characteristics like the slum tenements, are not characterized by precarious relationships of rental property.

Slum tenements are typical dwellings of rented rooms, with shared bathrooms. Some of them also have shared kitchens; others have no specific space for a kitchen but allow the dwellers to cook in the room itself. Slum tenements are also called rooming houses, boarding house rooms, or rental rooms[7]. In fact, it was found that most of the owners, administrators, and dwellers do not use the term "*cortiço*," slum tenement, perhaps by its pejorative nature in the popular imagination, since it is synonymous with precariousness, exploitation, poverty, and unhealthy conditions, bringing to mind the famous demolished slum tenements in the early twentieth century, the *Cabeça de Porco* (Pig's Head). Thus, in general, they use the terms "rooms" or "rented room."

7.2   A Little History: From the Demographic Explosion to the Attack Against the Slum Tenements

The population growth in Rio de Janeiro has considerably increased since the early nineteenth century. The arrival of the Portuguese court to the city in 1808 brought the disorders caused by the settlement of approximately 15,000 people—among noblemen, military, high-ranking employees, and the royal family—in an urban space that did not accommodate more than 50,000 inhabitants. Soon after, in 1810, the opening of the Brazilian ports to the friendly nations would intensify the trading movement in the capital, boosting a new migratory cycle. As a result, the population of Rio de Janeiro had practically doubled in less than two decades, reaching about 100,000 inhabitants in 1822, and reaching an impressive 135,000 in 1840 (Lamarão

---

[7] According to the *Dicionário Houaiss da Língua Portuguesa* (Editora Objetiva 2001), slum tenement is defined as "A house that serves as collective housing for the poor population; rooming house; pig's head."

2006). The figures continued to rise in the second half of the century. In 1888, the abolition of slavery triggered an exodus of freed slaves, especially from the coffee region of the State of Rio de Janeiro to the city, while European immigration was stimulated by the State to replace slave labor in the coffee plantations and "purify" the Brazilian "race" by the whitening of the population (Gonçalves 2013). Thus, between 1872 and 1890, the city saw a new demographic jump, from 266,000 to 522,000 inhabitants, and would still have to absorb the other 200,000 that arrived in the last decade of the century (de Carvalho 1987).

At that time, the urban fabric of Rio de Janeiro had already expanded to the suburbs of the South Zone of the city and the suburbs. However, it was still the central area that concentrated the jobs, bringing together not only trade and services but also much of the city's manufacturing park. Despite the expansion of public transport from the years 1860, transport costs accounted for a big share in the household budget for the vast majority. The labor market, on the one hand, and the precariousness of labor relations, on the other, forced the population to cluster in downtown neighborhoods, packed in slum tenements that multiplied. As a reflection of population explosion, slum tenement dwellers doubled in number between 1880 and 1890, totaling 100,000 people (Gonçalves 2013).

As it is known, the structure of the slum tenements was precarious. Owners and tenants earned extraordinary profit margins from a small investment, such as the construction of small houses or division of existing rooms in tiny rooms (Gonçalves 2013). The hygiene conditions of these overcrowded dwellings quickly deteriorated, transforming the slum tenements into outbreaks of sanitary infections and epidemics—such as the bubonic plague, yellow fever, and smallpox—recurring health issues in the city. Prainha, Saúde, and Gamboa—areas that not only contained slum tenements but were also the places where the port activities were carried out, with an intense flow of loads and people—were the epicenters of disease dissemination, a reason that led to the construction of Hospital Nossa Senhora da Saúde at the top of Morro da Saúde in 1853, and a bridge for the removal of corpses close to Cemitério do Caju (Caju Cemetery) on the Saúde coastal area, just behind the hospital in 1877 (Lamarão 2006).

Soon, the slum tenements, also associated with marginality, became the focus of the hygienist discourse which gained strength after the Proclamation of the Republic in 1889. Physicians and sanitarians affirmed that social factors were able to intensify the natural causes of diseases—such as the unhealthy conditions of the environment and malnutrition— so that it was imperative to fight both at the same time (Gonçalves 2013). In addition, the Republican regime was required to ensure the stability to the new power pact organized since the coup. In this scenario, the capital—politically mobilized, heterogeneous and socially fragmented, unruly, and divided by internal conflicts—represented danger, clearly perceived in military revolts, popular agitations, and labor strikes (de Carvalho 1987).

Thus, a control policy on the central space of the city was implemented, causing the banning of the lower classes from that area (Gonçalves 2013). This policy was based on three main elements: (a) restrictions on the operation, interdictions and gradual eradication of the slum tenements in the central area of the city; (b) criminalization of economic activities and cultural practices of the popular classes vital to their reproduction; and (c) encouragement to the construction of worker's houses, built in other parts of the city. Legislation has proved to be a fundamental instrument in this process. The imposition of legal restrictions on slum tenements has already been observed since the Empire: since 1856, the sanitary norms for the granting of building permits have become more severe, and, from 1880 onward, the *Junta Central de Higiene* [Central Hygiene Board] prohibited new constructions and demanded several collective houses to stop their operations (Gonçalves 2013).

However, compliance with the legislation has never been firmly applied to a large extent due to the slum tenement owners who, organized around the *Sociedade União dos Proprietários e Arrendatários de Prédios*, created to protect the rights of owners and tenants, have regularly used the Judiciary Branch and the Municipal Council to block the initiatives of the Central Hygiene Board (Gonçalves 2013). From 1889 onward, the Republican regime, relying on new social protagonists, reinforced the authoritarian nature of the hygienist measures and presented new reflections on the management of urban space, at the same time that it blocked the possibilities of participation of a large part of the population in political life (Gonçalves 2013; de Carvalho 1987). Thus, shortly after the installation of the provisional government, the Posture Code of 1890 was enacted, concentrating powers in the hands of the General Inspectorate of Hygiene, in addition to imposing sanitary requirements on buildings and expanding social control over the population living in inns and rooming houses (Gonçalves 2013). The 1891 Constitution, in turn, by requiring literacy from the voters, excluded the overwhelming majority of the population from voting, denying them their political right, and dissociated the municipal government from the representation of citizens, since the mayor, the position it created, would be appointed by the president of the Republic throughout the First Republic (de Carvalho 1987). As mayors were mostly physicians or engineers, they were often brought in from other states and were somehow disconnected from the life of the city. As stated by de Carvalho (1987, p. 35), "[...] it opened up then, on the government side, the path to authoritarianism, which at best could be an illustrious authoritarianism, based on the competence, real or presumed, of technicians."

Shortly thereafter, the policy of eradicating the slum tenements demolished the celebrated *Cabeça de Porco* (Pig's Head) which, according to reports from the newspapers of the time, housed about 4,000 people. The resistance of the slum tenements to the attempts made to eliminate it, during the Empire, had made the *Cabeça de Porco*—located near the Estrada de Ferro Central do Brasil (Central Railroad of Brazil), at the foot of Morro do Livramento—the most famous slum tenement in the city. It was rumored that its illustrious owner was count D'Eu, the husband of Princess Isabel. In 1891, however, the municipality signed a contract with engineer Carlos Sampaio who proposed to extend some streets and open a tunnel through the hill (the current João Ricardo tunnel) aiming to build properties and explore a railroad line (Cardoso et al. 1987). Amid the worsening of the epidemics in the early 1890s, Mayor Barata Ribeiro passed a decree on January 26, 1893, that allowed him to fight the slum tenements, and on that very day, he started to demolish *Cabeça de Porco* with an army of Public Hygiene employees, police, cavalry, municipal workers, and workers assigned by engineers such as Vieira Souto and Carlos Sampaio himself (Cardoso et al. 1987). At the end of the day, the slum tenement had vanished. The newspapers reported the demolition and announced the brief start of the works of the tunnel that would be completed only 30 years later when Sampaio was the mayor of the city (Vaz 1986).

Due to the social repercussions of the demolition of *Cabeça de Porco* and seeking to position himself before public opinion, Barata Ribeiro passed the Decree N°. 32, of January 29, 1893, expanding the benefits that had already been granted in the Empire, especially tax benefits, to produce working-class housing (Gonçalves 2013). The initiative contributed to develop the construction and real estate incorporation sectors; however, the number of houses built was insignificant, and the siege to the slum tenements became increasingly violent.

The legal texts were little by little increasing the perimeters that prohibited the installation of slum tenements. Thus, Decree N°. 762, of June 1, 1990, considerably increased the perimeter of the prohibition of collective houses, allowing their presence only in the neighborhoods of Gávea, Engenho Velho, São Cristóvão, Inhaúma, and Irajá (Gonçalves 2013). Two years later, Decree N°. 391 of February 10, 1903, passed when Pereira Passos had already assumed the municipal government, not only prohibited new constructions but also prohibited any construction work, renovation or repair that could favor the maintenance of the slum tenements, only allowing painting and whitewashing works (Gonçalves 2013). The management of Pereira Passos meant a new phase of resurgence against the slum tenements and the popular classes. Appointed by President Rodrigues Alves in 1902, Pereira Passos was entrusted with the task of cleaning up the port of Rio de Janeiro—its outdated infrastructure that imposed barriers to the growth of commercial exchanges, of beautifying the city and getting it rid of diseases, therefore transforming the capital based on the model of European cities. To this end, the President of the Republic endowed the new mayor with full powers, passing, on the day before the inauguration of Pereira Passos, a federal law that restructured the municipal administration, postponing the elections to the Municipal Council for 6 months (de Carvalho 1987). Without opposition, Pereira Passos used his discretionary powers to put into effect a cast of decrees aimed to facilitate the implementation of the "Passos Reform" construction works.

From 1903 to 1910, Rio experienced a radical transformation. On the one hand, the federal government was in charge of the works of the port improvements including: the landfill of Prainha, of Valongo and the entire coastline of the port area, extending the *Canal do Mangue* (Mangrove Channel) to the sea, and opening large roads for the movement of goods, as Central Avenue, Mangue Avenue and Cais Avenue (renamed, subsequently, Rio Branco Avenue, Francisco Bicalho Avenue, and Rodrigues Alves Avenue, respectively). On the other hand, the municipal administration concentrated its efforts in enlarging, extending, and opening new streets in densely populated areas—a project that became known as "Bota Abaixo" (Boot Down)—in addition to building squares and monuments such as the Municipal Theatre, the National Museum of Fine Arts and the National Library.

At the end of Mayor Pereira Passos management, in 1906, about 1,700 buildings had been demolished, and at least 20,000 people were removed (Gonçalves 2013).

The central administrative neighborhoods of Candelária, Santa Rita (where today the neighborhoods Saúde and Gamboa are) and of Sacramento (comprising the vicinity of Tiradentes Square, Saara, and the Red Cross) were the most sacrificed (Abreu 2013). But it was not just the ostensive demolitions that displaced the population: the valuation and speculation of urban soil, and its effects on rental prices; new taxes arising from the provision of new services, such as public lighting; the constraints and architectural parameters required for new constructions; and the prohibition of economic activities related to the livelihood of the popular classes acted as a powerful segregating force (Benchimol 1992). Thus, a spatial division was consolidated in the city of Rio de Janeiro, based on the hierarchization of the places. While the central area was reserved for business, the neighborhoods near the South Zone, located along the waterfront, were destined for the middle and upper classes. For the poor, the suburbs were only what was left.

In fact, a large part of the expelled population ended up moving to the suburban administrative neighborhoods closest to the central area such as Engenho Novo and Inhaúma (Abreu 2013). However, transportation costs and the high cost of construction materials made it difficult for workers to travel to the suburbs (Benchimol 1992). Many, then, ended up climbing the hills—especially the *Morro da Providência* that at the time had already been occupied by families of freed slaves and soldiers discharged from the Paraguayan War—giving rise to the favelas. It is in this sense that Vaz (1994, p. 592) says that "[…] the favela has at its origin the action of the same socio-spatial process that determined the end of the slum tenements." But many families still rented rooms. The slum tenements continued to exist, covering up the signs of their existence to survive the persecution and the numerous transformations and urban reorganizations the city has faced ever since.

7.3 Visibilizing the Slum Tenements in the Urban Landscape of Rio de Janeiro

Urban and social invisibility is a constitutive element of the history of the slum tenements in Rio de Janeiro and a key element to understand the dynamics that crosses and constitutes these spaces. This invisibility seems to be associated with what Wacquant calls, when discussing the case of racial segregation in the USA, a nexus between territorial stigma, lack of safety and abandonment by the State, making the dwellers of these areas the "city outcasts" (Wacquant 1995). The first expression of invisibility of this form of housing is in the total lack of information about the slum tenements in municipal public bodies. The city of Rio de Janeiro does not have any survey of the properties that rent rooms or of the socioeconomic profile of their dwellers.

Once the lack of public information was verified, the strategy for conducting the survey was to identify, street by street, the properties that operate as rental rooms. But during the work of identification, we met a first practical difficulty: most of the properties that operate as slum tenements are not identifiable by the facade, and do not have, in most cases, any type of board or advertisement identifying their purpose. In general, when real estate exhibits some identification, they are called hotel or lodging, even when they serve permanent housing for part of their "guests." As it can be observed in the course of this article, the question arises here in the fluidity between what is considered a permanent or a temporary dwelling by those living in the slum tenements.

Given the difficulty in identifying the properties based on their physical characteristics or advertisements, the fieldwork was organized in order to ensure that the teams would cover the overall streets of the area comprising Porto Maravilha. In general, the strategy used to identify the slum tenements was based on indications given by local and informal trade workers in the area and by slum tenement dwellers that had already been identified. Due to the turnover of some dwellers among the available slum tenements, many know how to point out other properties where it is possible to rent a room. In addition, in some cases, owners and/or administrators were responsible for more than one slum tenements in the area.

With this methodology, we were able to identify 54 slum tenements located in the port area of the Urban Operation, distributed in the neighborhoods of Santo Cristo, Gamboa, and Saúde, and also in Downtown streets, involving, according to the survey estimates, at least 712 rooms, where 1,120 persons dwell (Fig. 7.1). From the total universe of identified slum tenements, we were able to interview 25 administrators or property owners and collect information on the functioning of the establishments. In other cases, when it was possible, information was sought with neighbors, local traders, or dwellers. Simultaneously, aiming to set up the profile of the slum tenement dwellers, we needed to interview them. Thus, we had to return to the property after the first contact in which we explained the survey goals. We interviewed 105 dwellers

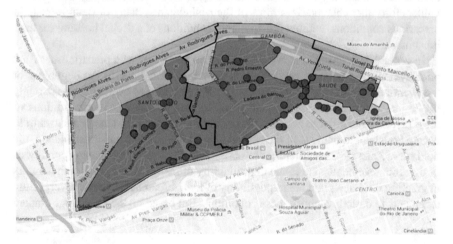

Fig. 7.1 Location of the slum tenements in the Porto Maravilha Urban Operation 2016. *Source* Observatório das Metrópoles 2016

living in the slum tenements identified in the port area, which can be considered a very significant sample of the total universe of the dwellers in these conditions. It is worth noting that it was not always possible to conduct interviews with the dwellers, either because of the difficulty of access to some locales controlled by the drug traffic or by the dwellers' mistrust, since the situation of vulnerability and informality of this type of housing also bring apprehension as there is no contract assuring them their permanence in such properties, a situation that is aggravated in cases in which the dweller is an immigrant who possibly lives illegally in the country.

During the field visits and interviews with the slum tenement dwellers, it was clear that there are information networks used by people to access this type of housing, in general, they are workers coming from other states of Brazil, sailors, immigrants—many of them in irregular situations in the country—and low-income families. Thus, networks of contacts between people who live or have already lived in slum tenements are formed, thus allowing information to be spread about what are the best alternatives to rental room in the central area of Rio de Janeiro, taking into account the needs and possibilities of those seeking this type of housing.

Still, regarding invisibility, the situation of immigrants reveals a particularity in relation to the other social groups. In many cases, the invisibility of the slum tenements can be used as a survival strategy. That is, for many immigrants living in an irregular situation in the country, the slum tenements and their social and urban invisibility are presented as a strategic means for those who are also, in a certain way, invisible in the country. As a result, the interviews were more difficult with this group.

But this invisibility has a price. In the first moment, we can understand the invisibility of the slum tenements in the urban landscape as a strategy to circumvent its illegality, allowing the reproduction of their operation. However, with the lack of laws governing and supervising the functioning of these properties, together with the vulnerability that characterizes most of their inhabitants, many of them operate in precarious conditions, without the minimum conditions of a decent housing for their inhabitants.

In general, it is common to associate the slum tenements with precariousness and poor hygiene and housing conditions, a view that dates back to the beginning of the twentieth century, as shown earlier. The survey aimed to overcome this strongly disseminated representation in society and to identify the concrete conditions of these dwellings in order to evaluate the possibilities of the slum tenements to shift to decent housing alternatives in the central area of the city. As previously stated, the field survey met several difficulties to obtain information on the housing conditions of the slum tenements, due to the unavailability of many owners and administrators in granting interviews and the impossibility of field researchers to enter several slum tenements to directly identify these conditions. But based on the interviews granted

Fig. 7.2 Facades of slum tenements identified in the port area 2016. *Source* Observatório das Metrópoles 2016

by 25 administrators or owners,[8] and also by the dwellers, it was possible to draw a very representative picture of the housing conditions of the slum tenements.

According to the interviewees, there are 712 rooms in the 54 identified slum tenements, and in two cases we were not able to obtain this information, which indicates that the total of rooms is still a little higher than that datum. Considering the universe of 52 tenements in which it was possible to collect information, it appears that in most cases (56%) they comprise small properties with 1–10 rooms (29 cases), but there is also a significant number (17 cases, corresponding to 33, 1%) of medium establishments with 11–25 rooms (Fig. 7.2). Large slum tenements, with more than 25 rooms, are a minority (only six cases, corresponding to 11%) but they include one with 60 bedrooms and another with 70 rooms.

---

[8]During the survey of the slum tenements, 25 owners or administrators of these establishments were interviewed. Among them, 12 were owners and 13 were administrators, and gender distribution was also balanced: among the owners, 6 were women and 6 were men; among the administrators, 7 were women and 6 were men. Almost all of them were Brazilians, except for one owner of English nationality. Among those who reported their place of birth, 10 were from Rio de Janeiro, 7 came from Northeastern states (3 from Ceará, 2 from Maranhão, 1 from Paraiba, and 1 from Rio Grande do Norte) and 1 was from Minas Gerais. The majority of those who reported their age, totaling 12 people, was in the age range between 30 and 59 years, while five others were between 60 and 75 years. Among those who reported their marital status, 11 were single, 5 married, 4 divorced, and 2 widowed. As in the issues involving property ownership, obscure in general, many stories are heard about the owners and administrators of the slum tenements involving cases where the real owners would be missing, the tenants sublet rooms, the drug traffic would control and manage some establishments, in addition to a case where a single policeman would control about 15 slum tenements in the central area of the city. As they were cases narrated by the dwellers themselves, it was not possible to check their truthfulness in the context of the survey.

The size of the rooms and the number of persons sharing each room seem to vary greatly, even in the same slum tenement, there can be rooms with different sizes. Considering the information of the 25 administrators and owners, complemented by information gathered from the dwellers, one can reach a reasonably precise picture in which it is clear that there are few slum tenements with rooms shared by more than two persons. In fact, it was found that in 22 slum tenements (42%) live only one person per room, while in other 23 slum tenements (44%) live up to two persons per room. There are seven slum tenements (14%) where the rooms are shared by more than two persons as follows: in five they were shared by three persons, one shared by five persons, and in another, the same room was shared by up to 12 persons. In the latter case, it is a slum tenement with a single room of more than 12 m^2.

Thus, considering the number of persons sharing each room, one would have a picture that would indicate housing conditions in the slum tenements that would not exactly correspond to their current social representation, marked by vulnerability, overcrowding, and bad living conditions. However, the situation becomes more complex when analyzing other fundamental conditions for acceptable habitability: the existence of windows, number and conditions of bathrooms, and availability of shared kitchens. Given the universe of 45 slum tenements where this information was collected, it is perceived that the number of slum tenements where all the rooms have windows, a requirement of the legislation, is very small, reaching only nine establishments (which represents 20% of the considered universe). In 13 other slum tenements (29%), most rooms have windows. In short, it can be noted that in most slum tenements, the conditions of the rooms are quite precarious since in 21 slum tenements (47%), the minority of rooms has windows while two (4%) have none.

Practically, all the slum tenements had shared bathrooms, except for two buildings that had bathrooms in the rooms,[9] and in some cases, it was possible to find rooms with or without individual bathrooms in the same slum tenement, varying the value of the room. From the information about the total number of bathrooms in each tenement, provided by 22 of the interviewed administrators and owners, a relationship can be traced between the number of the slum tenements dwellers and the available bathrooms. From this universe, nine slum tenements had between one and five dwellers per bathroom; six slum tenements had between six and 10 dwellers per bathroom; and seven slum tenements had between 11 and 17 dwellers per bathroom, thus being, at least apparently, the most severe cases of access to a good hygiene condition. However, during the field survey, 14 dwellers reported, from three different slum tenements, to live in properties with an average of more than 20 persons per bathroom, which may suggest that access to good hygiene conditions can be even more dramatic. But in addition to the number of shared bathrooms, the problems are also related to the infrastructure conditions of such facilities. The field

---

[9]The first image shows a small slum tenement of only five individual rooms; the second image shows a slum tenement whose rooms are shared by five persons. It can be said that bathrooms are individual only in the first slum tenement.

Fig. 7.3 Images of shared bathroom and kitchen of the slum tenements in the port area of Rio de Janeiro 2016. *Source* Observatório das Metrópoles 2016

survey revealed that the slum tenements have bathrooms in poor condition and that their vast majority has no hot water and, in some cases, no plumbing, as shown in Fig. 7.3.

As regards access to shared kitchens, the situation is also very precarious. Of the 51 tenements from where this information was gathered, we found that 28 tenements (55%) had shared kitchens, but a very significant number of 23 tenements (45%) did not, forcing their dwellers to dine outside the home (many reported eating in the popular restaurant of Central do Brazil), and/or use the room environment, with a small stove, with or without permission of the administrators, to make small meals. Although some dwellers recognized the risk of cooking in improvised facilities, they said that with such procedure they save money, which makes a difference at the end of the month and allows them to even send money to their families that often live in other cities or in other states of the country.

The invisibility of the slum tenements and their informal functioning, without being recognized by the municipal legislation, certainly feed the precarious situation observed in most of the buildings. But their dwellers still suffer from two more problems: the lack of contracts that give the minimum of security to their permanence in the rooms and the rental price that is relatively high in view of the living conditions offered.

Of the 25 administrators or owners who gave information on these topics, the vast majority, 20 of them, does not set any form of contract with the dwellers. The others affirmed to sign some type of a contract with their tenants which may range from 12 to 30 months. The lack of a contract makes the residence proof difficult. Some dwellers also reported having problems to open accounts in the bank, to obtain notary services, and even to enroll in social programs such as *Bolsa Família* (Family Allowance).

Finally, complementing this first framework on the housing conditions of the slum tenements, we attempted to evaluate the state of conservation of the properties based on the perception of the field researchers who should indicate whether such properties had bad conditions (with structural problems that may lead to risks to dwellers such as shored walls, compromised woods and improvised roofs), good (with no apparent structural problems) or medium (with mild structural problems, such as water infiltration, and exposed electrical wiring and pipes). In our survey, the sample corresponds to 48 properties where it was possible for the researchers to carry out this evaluation.

It is worth noting that only five slum tenements (11% of the total) were considered in good conservation conditions. But it is also worth highlighting that half of the properties visited, totaling 24 slum tenements (which accounts for 50%) were considered in a medium state of conservation, which means that with some investment in their infrastructure they could become a good decent housing option. In any way, this does not eliminate the severity of finding 19 slum tenements (39%) in poor housing conditions. The analysis of the conditions of the slum tenements indicates, contrary to the current perception, a picture of much heterogeneity, which could be simplified in the existence of two groups of slum tenements. A first group consisting of individual and shared rooms that have good infrastructure conditions and constitute an interesting housing alternative for some social groups that require a domicile in the central area of the city. And a second group consisting of rooms in a poor state of conservation that constitutes housing for several social groups due to the lack of a housing alternative in the central area of the city. It is important to point out that a large part of the slum tenements, perhaps most of them, is in an intermediate position between these two groups, with the potential to constitute a good housing alternative if reforms were made and regulations were instituted to ensure certain guarantees, including a stay contract and some control over rental price.

Unfortunately, however, the current situation, which feeds the vicious circle between invisibility, illegality, vulnerability, and precariousness, makes the space of the slum tenements a martyrdom, a violation of the right to housing for a large part of the population that needs to dwell and to live in the central area of the city.

7.4 Profile of the Slum Tenements Dwellers and Their Living Conditions

Who are the slum tenement dwellers? How do they live in these spaces? As previously observed when we analyzed the topic of housing conditions, there is a current social perception that associates dwellers with marginalization, linking them to the poorest strata of society. As it can be observed, the research reveals a heterogeneous social profile, of diversified social groups that demand access to the central area of the city.

It is worth noting that the research did not intend to make a census of the dwellers but to draw a general profile from a randomly defined sample, based on the availability of the dwellers in granting interviews. Evidently, this procedure poses some limitations, since possibly the survey could not incorporate dwellers who lived in irregular conditions, whether by their nationality (immigrants without permanent visas) or by their involvement with illegal activities, such as drug trafficking. Thus, it should be clear that this is the profile of a sample of slum tenements, but it reveals important information about this social universe.

The sample of this survey comprises 105 dwellers, representing about 9.4% of the total population estimated to reside in the slum tenements in the port area. Of the 105 interviewed dwellers, 77 were men and 28 were women, which may express the fact that men are a majority among those who rent the rooms, even because, as previously mentioned, many slum tenements rent room only for men. Of the 77 interviewed men, 51 were parents, but only four (about 10.5% of the sample) lived with their children. In the case of 28 women, 23 were mothers and most of them, totaling 13 women (72.2% of the sample) lived in the rooms with their children.

The condition of fatherhood and motherhood is reflected in the age range of the sample, consisting of relatively younger people than adults. A little more than half (54%) of the sample of the interviewed dwellers ranged from 30 to 59 years of age. The other part consisted of dwellers in the range from 16 to 29 years of age (27%) and in the range of more than 60 years (19%). Although the majority of 105 respondents were Brazilians, it was possible to incorporate in the sample of 10 persons with nationalities from other Latin American countries (4 Peruvians, 2 Argentinians, 2 Uruguayans, 1 Chilean, and 1 Venezuelan) and four from African countries (three Congolese and one Senegalese). But it is necessary to record the difficulty encountered to interview dwellers of other nationalities, given the insecurity regarding their permanence in the country.

Among 91 Brazilians, most of the sample was from outside the city of Rio de Janeiro. Apart from the 30% who declared themselves to be from the city itself, it is perceived the weight of the northeastern states, from which 47% of the dwellers came. Another 20% are declared natural from other states in the Southeast and only 1% of Southern states of the country.

An important dimension to understand the profile of the slum tenement dwellers relates to their current occupation. Based on the categories used by the *Observatório das Metrópoles* (de Queiroz Ribeiro and Ribeiro 2013), it is perceived that the majority of the dwellers (35% of the sample) has occupations linked to the nonspecialized tertiary, involving street vendors (most of the sample, with 16 persons in this occupation), nonspecialized service providers (10 persons), and domestic workers (6 persons). Another very expressive group is composed of inactive employees and retired people, comprising 21% of the sample (19 persons). In addition, the workers of the specialized tertiary (18% of the sample), involving trade workers (six persons) and specialized service providers (11 persons, 10 waiters or cooks) are highlighted; medium occupations (12% of the sample), involving artists and the like (nine persons) and office occupations (two persons); and workers from the secondary (11% of the sample), comprising auxiliary service workers (seven persons, five of whom are sailors), and civil construction workers (three persons). Finally, there is a small group of unemployed (3% of the sample, corresponding to three persons).

In large part, they are occupations whose work opportunities are strongly concentrated in the central area of the city, as in the case of street trading, work in kitchen and restaurants, trade in shops and sailors. But it also draws attention to the presence of retired people and inactive employees, who probably choose to live in slum tenements because they are in a well-structured area with good service offerings.

In addition to setting up the profile of the dwellers of the sample, we sought to characterize the living conditions in the slum tenements from some variables. Initially, it was sought to identify whether the slum tenements constituted a permanent dwelling for the dwellers themselves or if they considered their stay as provisional, which can also be confronted with the time they had lived in the rooms.

Generally, the slum tenements are considered by the dwellers as permanent types of housing, as stated by 98 of the 105 dwellers of the sample. Only seven dwellers claimed to be in the rooms provisionally, until finding a definitive solution or during a given period (as one of the dwellers who said they were living in the room only during the Olympics period and then would leave). But it also appears that the time of residence in the current room they are renting is quite varied, which may be explained by the great turnover in the different slum tenements since 53 dwellers (representing 52% of the sample) claimed that they had already lived previously in other slum tenements. In general, the central areas of the cities have better infrastructure, with electrical networks, water supply, and sewage collection already consolidated. The scenario found in the slum tenements seems to confirm this proposition: all 105 dwellers of the sample indicated that the properties have electricity and access to the water supply network. As previously seen, most rooms have no individual bathroom and kitchen, which implies that access to water is collectivized. However, this does not prevent problems in the provision of services, most likely due to the informality of the connections or the precariousness of existing infrastructures, as explained by 40% of the dwellers of the sample who said who suffer, or sometimes or often, problems in the water supply.

As it was already possible to observe, although the slum tenements share the same situation of informality, the conditions of conservation of the properties, as well as their infrastructure of services, vary greatly and this is reflected in the variation of rents paid by the dwellers, which confirms in a certain way the information provided by the owners and administrators. In view of the price ranges charged by the slum tenements, most dwellers pay intermediate values. Thus, it was identified that 59% of the dwellers of the sample (accounting for 60 persons) pay between R$ 301.00 and R$ 500.00, while 28% of the dwellers (28 persons) pay values in the lowest range, between R$ 150.00 and R$ 300.00 per month. In the highest value payment range, between R$ 501.00 and R$ 800.00, there were only 13% of the dwellers of the sample, i.e., the minority. The values of rents reflect several variables: the location in the central area, the conditions of conservation of the properties, the infrastructure of services, but also the informality of rental contracts. In fact, it appears that almost all the dwellers of the sample, corresponding to 94% of the persons, did not have rental contracts.

This framework of the living conditions of the slum tenement dwellers reinforces the idea of heterogeneity and diversity as the main characteristics of this social group, making it impossible and incorrect to construct an ideal type of the dweller of this type of housing. Within this social group, several subgroups can be distinguished.

Some are retired and inactive people, many are away from their families of origin and find in the space of the slum tenements not only the accessibility to the service network offered by the City Center, but also an affective community, a network of sociability that offers new possibilities for social reproduction. "*Seu*" Florindo, 70 years old, was in this situation: retired dancer and a single childless man, he has been living for 25 years in this slum tenement on Cunha Barbosa Street, in Gamboa, which rented nine individual rooms for singles. "*Seu*" Florindo proudly told his story as a black dancer, his travels around the world, and said: "I like this place, I feel welcomed."

Others are street vendors, informal workers, or workers without specialization, many living alone or away from their families, seeking to survive in the work of the streets of the central area of the city, who found in the slum tenements a housing alternative close to their area of work. A good example that illustrates the situation of this group is the story of Leandro, 21 years old, single. Two years ago, he started living in a slum tenement on Senador Pompeu Street, downtown, where he lived alone. Leandro came from the state of Espírito Santo, where he left a son, to work as a kitchen helper. As he reports: "I earn R$ 1,300.00 per month, I pay a rent of R$ 400.00, and send R$ 600.00 to my son, and I live with what is left. As soon as possible, I want to go back to my city."

Others are workers, neither so young nor elderly, but who still are in a stage of building their lives and consolidating their professional space, working in trade, as waiters or cooks, many are still single and find in the slum tenements a housing alternative consistent with their income and close to their potential labor market. The story of Katia is quite illustrative of this group. Katia is 34 years old, childless, and has been living in a slum tenement in João Homem Street, in the neighborhood of Saúde for 3 years, in an individual room. Natural from the Northeast, in her own words she says: "I intend to collect money to go back and buy a house in Maranhão; Rio is very violent." But there are also families, housekeepers, and workers in different nonspecialized branches, living with their children, husbands, and wives, in small rooms and without conditions to house a family, for whom to live in the slum tenements can be an experience of marginalization, stigma, and social exclusion. "*Dona*" Laura, 47 years old, was working as a street vendor, was married, and had been living for 2 years with her three children in the slum tenements of Senador Pompeu Street, in the central area of the city. She had already enrolled in the My House My Life program, but she said, "[…] never been called". As she said, "My dream is to get a bigger dwelling to house my family, even if it's in the West Zone, and even if I keep working at the Center."[10]

The reality of the slum tenements is complex not only because of the diverse set of housing conditions found among the rooms, many of them marked by precariousness and vulnerability, but also by the conditions of labor exploitation, the high costs and the precariousness of the mobility system in the city, which makes housing far from work a great issue. That is why the rent of rooms, even in precarious conditions, is a housing alternative in central areas for different social groups.

In this way, these different social groups that comprise the population inhabiting the slum tenements in Rio de Janeiro merge into the demand to live in the City Center. In fact, this type of housing is historically part of the landscape of several Brazilian and Latin American cities and remains one of the options the popular sectors have to reside in urban centers, even if living in conditions of vulnerability and conflict (Kowarick 2013; Toscani 2016). According to Lefebrve (2008) if, on the one hand, the urban phenomenon surpasses the old contradiction between the city and country, it causes the contradiction between center and periphery to emerge and it is from this contradiction that we can understand the issue of centrality for the author. For Lefebvre, the central regions would be those that gather power, culture, quality of life, and consumption, and are not necessarily limited to the geographic center. Thus, we can understand the need for access to centrality as a unifying demand of the different social groups that support the maintenance of the slum tenements in city centers. This can be well illustrated by the main motivations of the dwellers.

Their main motivations to live in this area of the city were its proximity to the Center and its good infrastructure, with 42% of the responses (46 persons) and they chose to live in the area because they worked there, with 41% (43 people). But also, they cited as motivations the fact of liking the area because of the people living in it, their history and their places, cited by 13% of the dwellers (14 persons), and the fact of identifying themselves with the area because they have relatives or friends in it, with 9% of the responses (nine persons). Finally, it draws attention the fact that only 10% of the dwellers (11 persons) said they were in the central area because the rent was cheap or because they had no other option, which indicates a high degree of satisfaction with the location of the house in the central area. This is also confirmed by the fact that 70% of the dwellers of the sample (71 persons) claimed that they liked to dwell in the slum tenements where they lived, while only 25% of the dwellers (26 persons) claimed they did not like, or hated, living there, even though there was a very small number, 5% of the dwellers (five persons) that liked it more or less.[11]

---

[10]The real names of the dwellers cited in this text have been altered to preserve their identities, and all other information was kept in the same way they reported it.

[11]For this response, the universe of the sample was 102 people, considering that 3 persons did not respond to that question.

7.5 Conclusion

The survey reveals that the slum tenements are a reality, constituting a housing alternative for several social groups in the port area, and should be visibilized and recognized by the municipal government through their regulation as it occurred in São Paulo, where specific legislation regulates the operation of rooms for rent.[12]

We sought to deconstruct the current perception that stigmatizes these spaces as informal, precarious and marginal, showing that the slum tenements are marked by a great heterogeneity of housing conditions and social groups, unified in their demand to live in the central area.

The recognition of this diversity of habitability conditions and the heterogeneity of social groups living in the slum tenements indicate the need for public policies that consider the plurality of situations. Families with children cannot live in one room and we need to think about housing alternatives that meet their needs for social reproduction with dignity. But for the other social groups identified here, the slum tenements can be a good alternative, provided that the quality requirements of this type of housing are established and guaranteed, with a minimum standard of habitability conditions, which include the minimum room size, windows requirement, limit on room sharing, regular and quality access to the water supply network, sewage collection and electric light, the functioning infrastructure of the slum tenements as bathrooms and kitchens in sufficient quantity and quality, rental contracts, and affordable prices for the low-income population.

The occupations of a large part of these slum tenement dwellers, linked to their housing conditions, also allow the triggering of some ideas formulated by Castel in his discussion about the "crisis of the wage society" (Castel 1998). To what extent the invisibility of the slum tenements would not feed the constitution of a "precarious periphery" within the centrality of Rio de Janeiro? Inspired by Castel (1998, p. 527), we could argue that the reproduction of this precarious periphery expressed in the slum tenements must be interpreted simultaneously from some processes: (i) "destabilization of the stable," (ii) "installation in precariousness," and (iii) "deficit of social and spatial places" offered to the popular classes.

---

[12]In 1991, the City of São Paulo enacted the Moura Law (Municipal Law N° 10.928, of January 8, 1991), whereby the slum tenements were recognized as a type of housing, establishing the minimum conditions for guaranteeing their habitability. With the Moura Law was instituted, the mandatory registration of the slum tenements in the municipality, which allowed the work of supervision by the city. From then on, other decrees and laws were published in order to regulate and apply the law: Decree N° 30.731, of November 12, 1991; Law N° 11.945/1995; Law Roberto Gouveia (State Law N° 9.142, of March 9, 1995). In general, the law of São Paulo recognized the organizations of the slum tenements dwellers, established rights and sanctions, and created financing mechanisms whereby the owners could make the necessary improvements in the real estate, so to meet the requirements of the law. Finally, the law of São Paulo also instituted interventions in slum tenements as a possibility in the context of its housing policy.

In any case, the set of public policies designed for the slum tenements should have central strategy the insurance of the right of these populations to the central area of the city. As it has been seen, what unifies the diversity of situations is the demand for centrality.

In view of the informality and invisibility of the slum tenements in the housing diagnosis of the port area and the lack of proposals in the Social Interest Housing Plan (PHIS) for the port area, it is also necessary to revise this plan, so that this type of housing is recognized and that proposals are incorporated to make the slum tenements a decent housing alternative in this area.

A policy designed for promoting the permanence of the popular classes in the central area of Rio de Janeiro will probably be undermined if the slum tenements are not recognized as an alternative type of housing. In fact, from the point of view of the promotion of social interest housing in the port area, the diagnosis point s to a failure of the Porto Maravilha Urban Operation after nearly 10 years of its implementation, considering that among its principles were the promotion of "social interest housing and attendance to the population residing in areas object of expropriation," "improvements in the areas of special social interest and its surroundings, with the implementation of infrastructure and land regularization"; and encouragement "to the recovery of occupied properties for the improvement of housing conditions of the resident population" (§ 1, art. 2 of the Complementary Law N°. 101, of November 23, 2009, which establishes the Urban Partnership Operation of the Port of Rio de Janeiro). However, so far, as Werneck notes (2017), "[…] there is no […] indication of different strategies for the feasibility of the actions foreseen in the Porto PHIS, not even studies of the scenery."

References

Abreu M (2013) Evolução urbana do Rio de Janeiro. Instituto Pereira Passos, Rio de Janeiro
Benchimol J (1992) *Pereira Passos*: um Haussmann tropical: a renovação urbana da cidade do Rio de Janeiro no início do século XX. Rio de Janeiro: Secretaria Municipal de Cultura, Turismo e Esportes, Departamento Geral de Documentação e Informação Cultural, Divisão de Editoração
Cardoso E, Vaz L, Albernaz MP, Aizen M, Pechman R (eds) (1987) História dos bairros: Saúde, Gamboa e Santo Cristo. Editora Index, Rio de Janeiro
Castel R (1998) *As metamorfoses da questão social*: uma crônica do salário. Vozes, Petrópolis (RJ)
de Carvalho JM (1987) *Os bestializados*: o Rio de Janeiro e a República que não foi. Companhia das Letras, São Paulo
de Queiroz Ribeiro LC, Ribeiro, M. G. (2013). *Análise social do território*: fundamentos teóricos e metodológicos. Letra Capital, Rio de Janeiro
de la Paz Toscani M (2016) Disputas en la ciudad: los procesos organizativos para resistir a los desalojos de los hoteles-pensión de la Ciudad de Buenos Aires. Article presented at III International Seminar of *Relateur Network*, La ciudad latinoamericana entre Globalización, Neoliberalismo y Adjetivaciones. Querétaro, México, 11–13 October 2016. http://observatoriodasmetropoles.net/images/abook_file/relateur3_toscani.pdf
Gonçalves RS (2013) *Favelas do Rio de Janeiro*: história e direito. Editora PUC-Rio, Pallas, Rio de Janeiro

Kowarick L (2013) Cortiços: a humilhação e a subalternidade. Tempo Social. São Paulo 25(2):49–77. https://dx.doi.org/10.1590/S0103-20702013000200004

Lamarão S (2006) *Dos trapiches ao porto*: um estudo sobre a área portuária do Rio de Janeiro. Secretaria Municipal de Culturas, Departamento Geral de Documentação e Informação Cultural, Rio de Janeiro

Lefebvre H (2008) Espaço e política. Editora UFMG, Belo Horizonte (MG)

Sasntos Júnior OAS, Lacerda L, Werneck M, Ribeiro B (2017) Invisibilidade, heterogeneidade e vulnerabilidade: os cortiços na área portuária do Rio de Janeiro. In: *Anais* XVII Encontro Nacional da Associação Nacional de Pós-Graduação e Pesquisa em Planejamento Urbano e Regional. ENANPUR, São Paulo

Saule Júnior N, de Almeida GMJA, Fontes MLP, de Menezes Cardoso P (2007) Possibilidades legais de proteção da moradia adequada nos cortiços. In: Saule Júnior N (ed) *Direito urbanístico*: vias jurídicas das políticas urbanas. safE—Sergio Antonio Fabris Editor, Porto Alegre (RS), pp 369–407

Vaz L (1986) Notas sobre o Cabeça de Porco. Revista Rio de Janeiro. Niterói 1(2):29–35

Vaz L (1994) Dos cortiços às favelas e aos edifícios de apartamentos—a modernização da moradia no Rio de Janeiro. Análise Social XXIX(127):581–597

Wacquant LJD (1995) Proscritos da Cidade: estigma e divisão social nos Estados Unidos e na França. Novos Estudos CEBRAP (43):64–83

Werneck M (2017) Habitação social no Porto Maravilha: cadê? Site Observatório das Metrópoles, 24 May 2017. http://observatoriodasmetropoles.net.br/wp/habitacao-social-do-porto-maravilha-cade/. Accessed August 2018

Chapter 8
Favelas and Gentrification: Reflections on the Impacts of Urban Restructuring on the City of Rio de Janeiro

Patricia Ramos Novaes

Abstract This article focuses on the urban, social, and symbolic transformations in Rio favelas, which became visible at the end of the first 2000s decade, linked mainly to public security, urbanization policies, and the arrival of the formal market in the favelas. These transformations began to be analyzed not only by the academic literature—but also by the media and local leadership—from the perspective of gentrification. In the city of Rio, research show that the adoption of the neoliberal pattern of urbanization in the early 2000s led to the elitization of popular territories. Thus, the concept of gentrification began to be applied in several studies on urban renewal in old port neighborhoods, industrial districts, and even in favelas, especially those located around the South Zone, where the elite is concentrated. However, from the years of 2015, attempts to resume drug trafficking in these territories and the economic crisis of Rio State have been undermining the public security program in favelas, placing barriers to the neoliberal urbanization of the city and consequently to gentrification experiments there.

Keywords Favela · Entrepreneurship · Neoliberalism

8.1 Introduction

Since the early 2000s, changes in the urban governance pattern of Rio de Janeiro have led to several socio-spatial adjustments and restructuring in the city justified by the preparation of the city to host major international events, notably the 2014 World Cup and the 2016 Olympic Games, and the alleged social legacy they would provide to the city.

Favelas, especially those located in the South Zone of the city, stood out as one of the territories where these adjustments and restructuring were implemented

P. R. Novaes (✉)
Observatório das Metrópoles, Institute of Urban and Regional Planing, Federal University of Rio de Janeiro, Rio de Janeiro, Brazil
e-mail: patricia.r.novaes@gmail.com

© The Editor(s) (if applicable) and The Author(s), under exclusive license   131
to Springer Nature Switzerland AG 2020
L. C. de Queiroz Ribeiro and F. Bignami (eds.), *The Legacy of Mega Events*,
The Latin American Studies Book Series,
https://doi.org/10.1007/978-3-030-55053-0_8

through a series of public investments in public security, urbanization, and local entrepreneurship.

The impact of these investments led to a process of symbolic resignification of these territories historically marked by the discourse of poverty, precariousness, and violence. Thus, new representations for the favela emerged through the terms "*Favela Chique*" [Chic Favela], "*Favela Modelo*" [Model Favela], and "Favela Criativa" [*Creative Favela*].

In addition, in these favelas, the emergence of new ventures aimed at a middle-class audience has been noted—a population that is settled outside the favelas— such as bars, hostels, bistros, and tourist guiding service. A process of real estate speculation has also been noted, with prices for buying and selling dwellings never before practiced in these territories. This impacted on the cost of living and led some dwellers to leave their dwellings or to relocate within the favelas, in areas not reached by speculation. Such transformations started to be analyzed not only by the academic literature, but also by the media and local leaders—as a process of gentrification.

After the period of the mega-sporting events, after 2016, the reversal of public investments in the favelas took place and, consequently, the reversal of the symbolic resignification and gentrification experiment process.

From this context, this article aims to discuss the impact of urban restructurings of the city in the context of mega-sporting events, in three favelas located in the South Zone of the City of Rio de Janeiro: Vidigal, Babilônia, and Chapéu Mangueira.

The article is divided into three sections. In the first, we discuss the relationship between the mega-sporting events and the process of urban restructuring in the city, marked by deepening of socio-spatial inequalities.

In the second, we discuss the concepts of gentrification and studies in the literature that have pointed out this phenomenon in the favela territories in the city.

In the third and last part, we present the three favelas, our focus of analysis, and then we analyze the processes of resignification and the gentrification experiments they underwent from 2010 to 2017, which led us to call this process peripheral gentrification (Novaes 2018).

8.2   Mega Events and Urban Restructuring in the City of Rio de Janeiro

From the 1990s onward, the dissemination of the neoliberal ideology in Brazil— headed by the articulation between international capital, segments of the national capital, and the State, under the leadership of the financial capital—brought changes to the management pattern of the cities.

This new management pattern is characterized, among other elements, by a coalition of interests based on the concept of public–private partnership (PPP) involving large national and international economic groups and the central core of local governments (municipalities and state governments), seeking and attracting external sources of funding as well as new direct investments. Thus, the city starts being ruled from the mercantile perspective, and urban planning starts being characterized as a competitive, flexible, market-friendly, and market-oriented strategic planning (Fleury and Ost 2013).

Following this approach, urban marketing gains prominence as it diagnoses and promotes the specific characteristics and attributes of each locality with the potential to attract large investments to implement speculative projects for cities.

In this context, the rise of neoliberalism and the diffusion of urban entrepreneurship would bring about several implications for the dynamics of cities in the direction of the so-called neoliberal urbanization.

According to Hackworth (2010), neoliberal urbanization has been increasingly characterized by a combination of appreciation of the central areas of cities, selective policies to stimulate development in certain areas of the city, relaxation of land use regulations, in addition to a reduction in public investments that do not generate immediate profits. Therefore, the author highlights, as icons of the neoliberal city, the restructuring projects of central areas, the production of mega events and urban megaprojects, and gentrification. In this sense, neoliberal urbanization presupposes both the destruction of urban structures and regulations (laws, decrees, norms) in force, and the creation of new structures, urban regulations, and forms of state management appropriate to the mercantilization of cities.

In the case of favelas, the absence of state regulation for years has led not only to the emergence of informality in the access to housing, electricity, water, etc., but also to the dominance of these territories by drug trafficking. Seeking to adjust this process, new regulations and forms of state management were designed for this space; it is worth understanding to what extent they are linked to the expansion of access to citizenship or are appropriate to the mercantilization of these territories, thus characterizing a process of neoliberal urbanization.

Since 1993, the city has had successive municipal administrations that have in some way pointed to this entrepreneurial nature. In 2007, Rio de Janeiro hosted the Pan American Games, certainly a smaller event, but one that already indicated an orientation of urban policy to the realization of major events. Its implementation was marked by problems such as extrapolation of the initial budget, conflicts regarding expropriation of dwellings, and private appropriation of urban infrastructure created primarily with public resources.

Between 2014 and 2016, the city hosted the World Cup and the Olympic Games which, according to Santos Junior (2015), were the expressions of an urban project to restructure the city, marked by the deepening of socio-spatial inequalities and experiments of the gentrification of three localities: the neighborhood of Barra da Tijuca, the port area, and the South Zone.

The municipal public power appears as one of the main promoters of these urban renewal projects, acting in various ways, involving the articulation or elaboration of projects, the direct financing of various interventions, the granting of tax incentives and tax exemptions for the attraction of private enterprises, the establishment of public–private partnerships, and the adoption of new institutional arrangements for the management of urban space and changes in the legislation previously in force, in particular, the legislation related to construction parameters.

Specifically, the ongoing transformations in the urban order of the South Zone of the city seem to suggest that this territory has experienced processes that combined the symbolic resignification of the favelas in this region, the real estate valuation of the areas not only located around the favela territories, but also the areas within these favelas, and the risk of gentrification. The favelas located in the South Zone of the city were the subject of three policies promoted by the public power: (i) public security, through the *Programa de Unidade Pacificadora* (PUC) [Pacifying Unit Program], (ii) urbanization, through the *PAC-Favela* and *Morar Carioca* programs, and (iii) incentive to local entrepreneurship.

However, after the cycle of the mega-sporting events—2014 World Cup and 2016 Olympic Games—it is important to know whether this resignification of the favelas in the South Zone was able to lead these territories to gentrification processes.

8.3 Favelas and Gentrification

Based on the formulations of Neil Smith (1987, 2006), the concept of gentrification implies the process of restructuring the central areas—in decline and occupied by the low-income population—through the action of public and private collective actors (entrepreneurs, real estate, banks, public managers, individual owners) moved by both the location characteristics and the less valued land price compared with other areas of the city. Thus, renovation of houses or even new constructions for the middle class, besides the establishment of new ventures and services in these areas, also attract new dwellers, which gradually causes the departure of the old dwellers by the increase in the cost of living and the decharacterization of the location.

The first debates about the phenomenon of gentrification in Brazil, as well as in other Latin American countries, emerged from the 2000s (Janoschka et al. 2014) were related to the adherence of the state and municipal governments to "renewal" and urban revitalization policies. In many Brazilian cities, these policies have been priming for beautification, housing reordering, and creation of new spaces of culture and leisure addressed to the middle and high classes. It was noted that some popular territories in Brazilian cities suffered elitization and, therefore, in the following decade the concept of gentrification started to be triggered not only by researchers in their academic analyses, but also by the media and popular movements that felt the effects of elitization on the spaces where they lived.

Many studies have pointed out that the concept of gentrification has been precisely expanded to account for new experiences that the classical theory of gentrification could not explain. Among them, the experiences most linked to the transformations in the pattern of leisure, tourism and entertainment in opposition to those linked to residential standard, in which the middle-class population, to a greater extent, is more attracted to consumption than to housing, as pointed out in the literature on commercial gentrification (Charbol et al. 2014) and tourism gentrification (Mendes 2017; Gant 2015).

In the City of Rio de Janeiro, the literature relates the phenomenon of gentrification essentially to three localities: the favelas of the South Zone, the port area, and the neighborhood of Barra da Tijuca (Santos Junior 2015). Specifically about gentrification in the South Zone favelas, more recent research has pointed to changes in the pattern of commerce and services, in the profile of the population that circulates and consumes these services, besides the increased cost of living and new opportunities created for sale and rent of property for values never practiced in the favela (Ost 2012; Lacerda 2016).

It was also observed that the sites where the funk dances previously occurred started to be occupied by other sounds such as soul music, samba, and jazz, charging admission tickets with prices inaccessible to the dwellers (Teixeira and Fortunato 2016).

Other analyses have highlighted that the distribution of property bonds, resulting from land regularization in some favelas in the South Zone, has generated processes of privatization of public land stocks. This fact could progress toward the legitimation of gentrification processes, as it created new market conditions in territories that have been undergoing a strong process of economic, social, and cultural requalification (Bonamichi 2016).

In fact, the introduction of market dynamics with the absence of mechanisms of state regulations led to real estate speculation and the increase in the values of sale and rent practiced in the favelas. As a result, objective and subjective processes were generated increasing the cost of living and leading to the displacement of dwellers to other parts of the city, and even into other areas within the favelas. Besides, this market dynamics led to changes in the pattern of trade and services in these territories. In addition to the emergence of spaces addressing to tourism and middle-class consumption, there was an increase in the values practiced in some traditional trades that sought to adapt to the new demands.

However, what seems to us more relevant so far in the studies on gentrification in the favelas in the South Zone of Rio de Janeiro is that this phenomenon has not been completely established and is still an ongoing process. This is due to a number of reasons—such as the economic crisis that the country and the City of Rio de Janeiro began to face after 2015—which worsened with the end of the mega event cycle. Other obstacles seem to derive from (a) the symbolic weight of the favelas as territories of illegality, violence, and marginality, (b) the fragility of security and urbanization public policies, and (c) the local mobilization that sought to face the social and territorial exclusion processes in the favelas.

In this sense, we will apply the concept of peripheral gentrification (Novaes 2018) to interpret the experiences of elitization undergone by the favelas of the South Zone of the city. According to it, favelas are popular territories stigmatized by marginality and some of their spaces are undergoing the process of elitization. This occurs mainly in terms of the pattern of services and trades offered, thus reinforcing the dynamics of exclusion on the microscale and reproducing the unequal and combined socio-spatial structure of formality and informality of the City of Rio de Janeiro. In the

Fig. 8.1 Localization of the Vidigal, Babilônia and Chapéu Mangueira favelas. *Source* Observatório das Metrópoles 2019

next section, we will present the case studies in three favelas located in the South Zone of the city, seeking to explore the idea of peripheral gentrification.

8.4 Symbolic Resignification and Gentrification Experiments in the Favelas of Vidigal, Babilônia, and Chapéu Mangueira

With about 4,585 households and 12,797 dwellers,[1] Vidigal is located in the neighborhoods of Leblon. Its occupation dates from the 1940s, driven by both the urbanization of the neighborhoods of Ipanema and Leblon.

The *Chapéu Mangueira* and *Babilônia* favelas are located on the hillside of *Morro da Babilônia* in the neighborhood of Leme (Fig. 8.1). The occupation of its slopes started in the early twentieth century and was most prominent from 1930 onward when the two favelas were actually created. Currently, the *Babilônia* Favela has

---

[1]Rio de Janeiro: IBGE 2010. Synthesis of social indicators: an analysis of the living conditions of the Brazilian population. Available at: www.ibge.gov.br/home/estatistica/populacao/condicaod evida/indicadoresminimos/sinteseindicsociais2010/SIS_2010.pdf. Accessed Nov 12, 2016.

about 777 households and 2,451 dwellers and *Chapéu Mangueira* has about 401 households and 1,288 dwellers.[2]

Both Vidigal and Babilônia have trails that are a great tourist attraction. They lead to the top of *Morro Dois Irmãos* (Vidigal) and *Pedra do Urubu* (Babilônia) with beautiful views of the city. But the view and location are not the only reasons why these favelas became known. Their stories of construction and consolidation are crossed by moments that marked these territories as places of cultural effervescence, assuring them a certain escape route against the hegemonic discourses that marked the favelas as sites of illegality, marginality, and violence (Valladares 2005; Gonçalves 2013; Leite 2015).

These historical and cultural particularities are some of the fundamental points for understanding the socio-spatial transformations that these favelas have experienced, allied to the real estate valuation of this territory, especially after the implementation of the *Unidades de Polícia Pacificadora* (UPPs) [Pacifying Police Units].

After several attempts of the expulsion of their dwellers in the 1970s, Vidigal gained visibility and broader support from society in this period since the real reason for the eviction came to light: the company Rio Towers would have bought the area, where the favela is located until today, aiming to build a luxury hotel. The case caused impact; attracted the attention of lawyers, social movements, and artists; and culminated in a show to support the dwellers' fight. Subsequently, this process was further promoted by the creation of the *Teatro Grupo Nós do Morro*, in 1986, which also reached a considerable level of recognition and played an important role in the dissemination of the image of Vidigal as a place of active cultural production in the city.

Babilônia and Chapéu Mangueira have attracted the attention of film directors and served as a set for some films such as: *Orfeu Negro* (1959), by Marcel Camus; *Tropa de Elite* (2007), by José Padilha; *Chapéu Mangueira e Babilônia—histórias do morro* (1999) by Consuelo Lins; and *Babilônia 2000* (1999), by Eduardo Coutinho. More recently, in 2015, they were scenarios for a soap opera of the largest Brazilian television network, *Rede Globo*.

8.4.1 The "Pacification" Process of Favelas

From the 1990s, the favelas in Rio were marked by violence and criminality due to the significant increase in drug trafficking. According to Leite (2015), it is from that moment that favelas were envisaged, above all, as "territories of violence". Seen in the social perception as "territories of illegality," favelas are managed as "territories of violence". That is, this new representation of the favelas and their dwellers, linked to the image of violent drug trafficking, is superimposed on the representation of

---

[2]Rio de Janeiro: IBGE 2010. Synthesis of social indicators: an analysis of the living conditions of the Brazilian population. Available at: www.ibge.gov.br/home/estatistica/populacao/condicaod evida/indicadoresminimos/sinteseindicsociais2010/SIS_2010.pdf. Accessed: Nov. 12, 2016.

illegal spaces and creates a radically opposite image between favela and formal neighborhoods.

However, from the second half of the 2000s, in a period in which the image of the city emerges as its main attraction in the global market, violence becomes a problem to be fought. Thus, the public security program of the pacifying police units, launched in 2008, appears as a strategy for Rio de Janeiro to ensure the attraction of mega-sporting events and the opportunity for investments, speculation, and real estate appreciation on several fronts of the city. Azevedo and Faulhaber (2015) draw attention to the distribution of the UPPs across the territory and concentrated in the areas of greater investments to receive the mega-sporting events of the 2014 World Cup and 2016 Olympics. This indicates that the distribution of the UPPs did not follow a public safety logic in the city as a whole but rather the security of certain spaces for sporting events.

Along with the implementation of the security policy, the *Rio+Social*[3] program was established with the aim of integrating the favela in the city through access to public goods and services with coverage and quality compatible with those offered in the formal city.

It is undeniable that the "pacifying" police program has generated impacts on the city's violence rates. The sense of security has increased in view of the decrease in the ostensible use of weapons by trafficking and the reduction of shootings between rivaling drug gangs and between them and the police.

According to the report on the indicators of this policy carried out by the *Instituto de Segurança Pública* (ISP)[4] [Institute of Public Security], from 2009 onward, the first year after the UPPs onset, the rate of homicides decreased in the City of Rio de Janeiro, increasing the distance between the state rates. In 2014, while in the State of Rio de Janeiro the rate was 30 victims of first-degree murder, in the city, the rate was 19 victims. In addition, there was a reduction in the rate of violent lethality in the city, according to the graphs (Figs. 8.2 and 8.3).

With the onset of the UPPs, in addition to public security, there was a large investment in the regulation of services previously provided informally, such as electricity, water, and cable TV system, and the regulation of local transport and formalization of the local economy through the *Serviço Brasileiro de Apoio à Micro e Pequenas Empresas* (SEBRAE) [Brazilian service for micro and small business support].

Another service that was regulated and controlled by the UPP was transport by mototaxi and Kombi, widely used in the favelas which, in general, had been previously organized by drug-trafficking or militia groups.

With the Brazilian economic growth between 2003 and 2014, which increased the consumption power of popular classes, private companies had their attention drawn to the existing potential for consumption in the favelas. Thus, the favelas with UPP have become investment points of some brands for the commercialization of

---

[3]Initially called Social UPP, the name was changed in 2012 when the program was restructured.

[4]For further information about the report see: http://arquivos.proderj.rj.gov.br/isp_imagens/Upl oads/RelatorioUPPvfinal.pdf. Accessed: July 2016.

Fig. 8.2 Violent death rate in the Municipality of Rio de Janeiro. *Onset of the UPPs facilities. Victims of first-degree murders per 100,000 inhabitants. *Source* ISP 2014

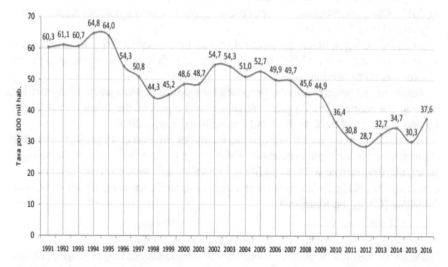

Fig. 8.3 First-degree murder rates in the State and in the City of Rio de Janeiro. Rate per 100,000 inhabitants. *Source* ISP 2016

products such as telephony and internet companies, cable TV operators, banks, and insurance companies. Such companies, in addition to promoting their brands as a result of social responsibility actions, also sought to attract new consumers.

Moreover, incentives for entrepreneurship were set up, especially through partnerships with SEBRAE, which conducted diagnoses in some favelas to provide guidance on business management (increased efficiency, productivity, sustainability, stimulus to business innovation), associativism, and access to credit.[5]

This partnership between UPP and SEBRAE encouraged the formalization of favela entrepreneurs through the issuance of *Cadastro Nacional de Pessoa Jurídica* (CNPJ) [National Register of Legal Entities], guidance for environmental and sanitary licenses as well as registration for training courses with tips of management and business expansion.[6]

Besides, between 2008 and 2012, a period of the onset of PPPs in the favelas of the South Zone, the press reported that real estate market entities pointed to a valuation of the properties in the area which reached 70%–100% in some areas of the neighborhoods (Folha de s. Paulo online 2011).[7]

8.4.1.1   The *Morar Carioca* Program and Urbanization in the Favelas

In 2010, the City of Rio de Janeiro designed the *Morar Carioca*, a large municipal program of urbanization of favelas that would be part of the *Plano de Legados dos Jogos Olímpicos*[8] [Legacy Plan of the Olympic Games].

The *Morar Carioca* initially predicted the urbanization of the overall favelas of the city until 2020, with the investment forecast of R$ 8 billion, with R$ 2 billion up to 2012. Thus, the program was set up as a major brand of the social legacy of the Olympic Games in Rio de Janeiro.

In addition, it predicted the removal of 123 favelas up to 2012, in which there were supposedly about 12,973 families in the landslide and flood-prone areas or environmental protection areas.[9]

However, the projects initially proposed for the favelas were just partially carried out and the *Morar Carioca* program ceased to be part of the Legacy Plan of the Olympic Games of the City Hall and ceased to exist after 2016.

In Babilônia and Chapéu Mangueira, the major infrastructure projects were the reformulation of the main avenue that gives access to the two favelas, to the sewage, water, and drainage network. In addition, in Babilônia,[10] were built two new squares, a multisport court and a belvedere that looks out upon Copacabana Beach and *Cristo Redentor* [Christ the Redeemer].

---

[5]Information retrieved from the website www.sebrae.com.br. Accessed: July 2016.

[6]Information retrieved from the website http://www.riomaissocial.org/tag/empresa-bacana. Accessed: July, 2016.

[7]Available at: http://g1.globo.com/rio-de-janeiro/noticia/2011/11/mercado-ve-imoveismais-caros-em-sao-conrado-com-upp-da-rocinha.html. Accessed: Mar. 14, 2016.

[8]Government commitments plan in partnership with the Olympic Public Authority and the 2016 Rio Committee to ensure benefits to the population with resources that had been provided for the Olympic Games.

[9]For further information, see: https://oglobo.globo.com/rio/prefeitura-lanca-novo-plano-para-fav elas-que-preve-controle-gabarito-conservacao-choque-2974915. Accessed: July, 2017.

[10]Information retrieved from the website of the *Morar Carioca* Program: https://medium.com/ morar-carioca/onde-o-morar-carioca-chegou-centro-5fb0572a058b. Accessed: Oct., 2017.

Some dwellings were demarcated for removal because they were in an area of landslide risk or in an environmental preservation area. In Babilônia and Chapéu Mangueira, the relocation of 144 dwellings was planned with about 400 people. Thus, three buildings were designed for the favelas—two in Babilônia and one in Chapéu Mangueira—totaling 116 housing units but only two buildings were built.

Hence, 28 housing units would not be built within the favela. According to the president of the *Associação de Moradores da Babilônia* [Babilônia Dwellers Association], the options given by the City Hall were resettlement in the North Zone and in the neighborhood of Santa Cruz, in the West Zone. However, the two options are more than 35 km away from the favela, which led to the strong mobilization of dwellers for their stay on the site.

Vidigal was not contemplated with any urbanization work from the *Morar Carioca* program; however, important interventions were made in the favela through the GEO-RIO, body of the *Secretaria Municipal de Obras* [Municipal Secretariat of Public Works] responsible for slope stability. During this period, two projects were implemented: the first made feasible with municipality resources, and the second, with the *Programa de Aceleração do Crescimento* (PAC) [Growth Acceleration Program] resources, totaling R\$ 6.1 million in investment.

In general, the projects were developed in two areas of Vidigal—known as Arvrão and Carlos Duque. These two areas are located in the upper parts of the hill, guaranteeing access to the view of part of the beaches of the South Zone of the city.

In Arvrão, the works were completed, and the area is no longer a risk area. Currently, it is the most coveted area in the favela where some of the new ventures are concentrated, such as the hostels and bars where parties often occur organized by known producers and DJs of the *Carioca* night. There are also other small trades that revolve around the public attracted by the new ventures.

In Carlos Duque, the project included the removal of about 40 dwellers but, due to their resistance, up to now, the removal has not been carried out. No public work was performed, and the region remains classified as a risk area. In interviews with the geologist responsible for the study of the area, the resistance of the dwellers would have impacted the realization of the works since the contract with *Caixa Econômica Federal* (CEF) [Federal Savings Bank], the resource manager, linked the effective removal to the release of this resource.

Certainly, the set of works conducted in the studied favelas contributed to the improvement of the living conditions of the dwellers. However, it can be noted that when the project requires housing removal and the dwellers resist this action, the area does not receive improvement works.

On the other hand, even with the works carried out, there are still numerous problems related to sanitation, garbage collection, and paving of the streets that give access to the areas at the top of the favelas. Moreover, the investments did not meet the needs resulting from the increased movement of people and transport in the favelas that have been made possible due to the public security program. It is worth noting that the flow of vehicles, formed in the main streets, does not hold the quantity of delivery trucks, cars, motorcycles, vans, and kombis circulating in the favelas. This, incidentally, was the agenda of numerous UPP meetings with dwellers and external entrepreneurs for the regulation of traffic and parking in the streets.

This question is especially important because it is a possible concrete obstacle to the advancement of the territory mercantilization of the favelas, since the physical barriers that prevent the movement of people and goods, coupled with incomplete urbanization, can bring losses or demand greater investment from the entrepreneurs.

8.5   Entrepreneurship Reaches the Favelas

Entrepreneurship, such as productive inclusion and generation of employment and income, became more widespread in Brazil from the 1990s as an unfolding of the neoliberal ideology that preaches economic liberalism and the breakdown of social regulation. In the face of restructuring in the global productive sector, structural unemployment on a global scale and a subproletarianization occurred, with expansion of partial, temporary, and outsourced work (Antunes 1999). This process imposed a new model worker, one who would adapt to new market structures and meet their requirements with flexible characteristics, one who would be polyvalent, creative, able to create new businesses, and provide to himself, regardless of market regulations.

Thus, in view of the growth of poverty and unemployment, entrepreneurial practices started to enter the agenda of public policies and nongovernmental organizations.

In the favelas in Rio de Janeiro, the dissemination of entrepreneurship, which has been carried out by NGOs since the 1990s, grew with the encouragement of the public power from the "pacifying" police and the *Rio+Social* programs. In addition, projects in partnership with the private sector, such as SEBRAE, were instrumental in stimulating entrepreneurship.

In addition, the *Agência Estadual de Fomento* (AgeRio) [State Agency for Development] was created by the state government to encourage the projects in the favelas with UPP. From 2013 onward, the agency started to grant financing between R$ 300.00 and R$ 15,000.00 with interest from 0.25% per month, through the UPP Entrepreneur Fund.

As a result of public policies and incentives, the favelas have become an opportunity not only for entrepreneurs living there, but also for external entrepreneurs who saw in these spaces the possibility of opening their business in the South Zone, making the most of the potential of this area of the city (such as nature, tourism flow, concentration of cultural equipment), with low cost in the acquisition or rent of the property. In this process, a new type of trade emerged, one that neither existed in these territories nor was accessible to the dwellers.

In the favelas of Babilônia and Mangueira, after the "pacification" in 2009, 14 hostels and 4 bars addressed to the public outside the favela were built. In addition, tourism circulation in the favela was favored by the increased guiding services to the trail of Babilônia, performed by local guides and external guides as well.

In 2010, right after the "pacification", the first favela hostel in the city, the Favela Inn, opened in Chapéu Mangueira. After this experience, three other hostels emerged in Chapéu Mangueira with external entrepreneurs' owners.

In Babilônia, all hostels were opened by external entrepreneurs, such as the *Aquarela do Leme* hostel (Fig. 8.4, owned by a resident of the neighborhood of Leme.

In the Vidigal favela, there were nine lodging venues: we point out the luxury hotel *Mirante do Arvrão* (Fig. 8.5), a venture designed by the famous architect Hélio

Fig. 8.4 Suite of hostel and inn Aquarela do Leme—Babilônia. *Source* Retrieved from the Internet/Aquarela do Leme. Available at: http://www.aquareladoleme.com. Accessed: June 30, 2018

Fig. 8.5 Suite of Hotel Mirante do Arvrão—Vidigal. *Source* www.tripadvisor.com

Fig. 8.6 Bar Da Laje—Vidigal. *Source* www.bardalaje.com

Pellegrino, and the hostel *Alto Vidigal*, of the Austrian Andreas Wielend. The Hotel *Mirante do Arvrão* has 14 suites and a collective room with daily rates that reach R\$ 600.00 in the main suite. According to the hotel manager, the rates are high because they are "selling an experience", that is, the opportunity of knowing one of the most beautiful favelas in the city. Much of the hotel's advertising comes indirectly since the bar space is constantly rented by producers who organize parties or recordings and photo shoots of artists for advertisements. This also explains the constant presence of famous people, which contributes to the dissemination of this 'cool' image of the hotel and also of the favela.

The same idea is applied to one of the most sought-after bars in the city, the *Da Laje* bar (Fig. 8.6), whose owner is an entrepreneur of artists of *Rede Globo* (Globo Television Network). On the menu, the type of the dish served, and its high cost, ensures a selected clientele to the bar.

The term favela seems to have been resignified in such a way that it is no more associated with the idea of a territory of deficiencies, but with the idea of an experience to be consumed. The term "favela" (Fig. 8.7) was in the names of the festivals promoted at the hotel *Mirante do Arvrão* in Vidigal, in the t-shirts sold by big brands and in the advertising of the *Bar do David* in Chapéu Mangueira (Fig. 8.8), winner of the regional and national contest of *Comida de Buteco*.

This whole process seems to reveal that favelas have undergone gentrification experiments. However, some blockages seem to change this process. Among them, there would be the symbolic weight of the favelas—identified from the first formations as territories of illegality, violence, and social vulnerability—and the fragility of the security and urbanization policies for the favelas.

In relation to public security, even with a pacifying police project, with explicit goals of bringing "peace" to these localities, the logic of management of these territories is still the logic of a repressive social control of the dwellers. Research (Abramovay and Garcia Castro 2014; Carvalho 2013; Leite 2012) has pointed out that in the perception of the dwellers of some favelas with UPP, the improvement in life

Fig. 8.7 The Osklen brand ad: "Vidigal" t-shirt. *Source* Retrieved from the Internet. Figure retrieved from the website: https://www.google.com.br/search?q=camiseta+osklen+vidigal&sourc. Accessed: June 30, 2018

condition occurred due to the reduction of ostensive weapons and the confrontation of drug traffickers with the police, but countless testimonies point to the persistence of abusive and violent police practices (vexing inspections, intimidation of dwellers, physical violence, etc.).

The abuse of power and the violence of the police in the UPPs do not seem to have disrupted effectively, the security policy practiced in a "war metaphor" context (Leite 2012). Thus, it can be said that the police treatment given to the favelas' dwellers, the *favelados*, has not substantially changed in quality, which can be explained by the patterns of control of the favela and the *favelados*, so crystallized in the police institution. The police actions directed to the favela territories historically reinforce the linkages between poverty and crime by identifying dwellers as potentially criminals.

In addition, from 2016 onward, the backlash in public security financing has been noted. Attempts to resume drug trafficking in these territories and the numerous clashes with the police and other rivaling drug gangs, coupled with the economic crisis that the State of Rio de Janeiro has been experiencing, have shaken both the program of pacification of favelas and the project of resignification and entrepreneurial practices in these territories. Repression processes within the favelas will possibly expand and trigger either by the police itself or by drug trafficking.

In relation to urbanization and housing policies, there is a reconfiguration of the programs. The *Morar Carioca* program was reduced and its main goal, favela urbanization, was not reached.

Fig. 8.8 Favela Chic. *Bar do David* ad in Chapéu Mangueira. *Source* Retrieved from the Internet. Figure retrieved from the website: https://pt.foursquare.com/v/bar-do-david/4daa3f23cda1652a2b 896ee2/photos. Accessed: June 30, 2018

In this context, we would rather state that the attempt at elitization in these favela territories seems to have been a neoliberal experiment that faced barriers and was not completely established, which can be understood if we apply the concept of peripheral gentrification in our analysis.

We have applied this concept to characterize a process that combines urbanization, public security policies, and commercial and tourism dynamics in a popular territory in an attempt to resignify, value, and integrate this area with the formal logic of neighborhoods in the South Zone. However, the appropriation by the middle class seems to have been verified in certain areas of the favela (not in its totality), which can lead to an internal differentiation and reproduction of the pattern of residential segregation that combines residential proximity and social distance of the classes.

8.6   Conclusion

The processes of urban transformations of these favelas seem more complex since, at least so far, it seems very unlikely to bet on the disappearance of the popular classes from these territories. They are still considered spaces of popular housing, but there is clearly a process of elitization in certain spaces where are concentrated hostels, parties, bars, and restaurants addressed to the middle class. For these reasons, we find it relevant to trigger the idea of peripheral gentrification to interpret the experiences of elitization that the favelas of the South Zone of the city have been undergoing.

These socio-spatial transformations experienced by the favelas are elements in dispute and depend on the correlation of forces of the actors involved. Currently, attempts to resume drug trafficking in these territories, coupled with the crisis that the State of Rio de Janeiro has been undergoing after the cycle of the mega-sporting events, are affecting the pacification program of favelas and, consequently, placing another barrier to the attempt of elitization of favelas in the South Zone of the city.

References

Antunes R (1999) Os Sentidos do Trabalho. São Paulo: Boitempo

Azevedo L, Faulhaber L (2015) *SMH 2016*: remoções no Rio de Janeiro Olímpico. Mórula, Rio de Janeiro

Bonamichi NC (2016) Favela on sale: regularização fundiária e gentrificação de favelas no Rio de Janeiro. Master's thesis. Universidade Federal do Rio de Janeiro, Instituto de Pesquisa e Planejamento Urbano e Regional, Programa de Pós-Graduação em Planejamento Urbano e Regional

Carvalho MB (2013) A política de pacificação de favelas e as contradições para a produção de uma cidade segura. O Social em Questão - Ano XVI - nº 29

Castro et al (2014) O Projeto Olímpico da Cidade do Rio de Janeiro: Reflexões sobre os impactos dos megaeventos esportivos na perspectiva do direito à cidade. In Rio de Janeiro: Os impactos da copa do mundo 2014 e das olimpíadas 2016, Letra Capital

Charbol M, Fleury A, Van Criekingen M (2014) Commerce et gentrification. Le commerce comme marqueur, vecteur ou frein de la gentrification. Regards croisés à Berlin, Bruxelles et Paris. In: Gasnier A, Lemarchand N Le commerce dans tous ses états. Espaces marchands et enjeux de société. Rennes, PUR

Fleury S, OST S (2013) O mercado sobe o morro: a cidadania desce? Efeitos socioeconômicos da pacificação no Santa Marta. Dados 56(3)

Gant A (2015) RC21 International Conference on "The Ideal City: between myth and reality. Representations, policies, contradictions and challenges for tomorrow's urban life" Urbino (Italy)

Gonçalves R (2013) Favelas do Rio de Janeiro. História e direito. Pallas: Editora PUC-Rio, Rio de Janeiro

Hackworth J (2010) The Neoliberal City: Governance, Ideology and Development in American Urbanism. New York: Cornell University Press

Janoschka M, Sequera J, Salinas L (2014) Gentrification in Spain and Latin America — a Critical Dialogue. Int J Urban Regional Res 38(4)

Lacerda LG (2016) Conflitos e disputas pela mercantilização de territórios populares: o caso da favela do Vidigal. Master's thesis. Universidade Federal do Rio de Janeiro, Instituto de Pesquisa e Planejamento Urbano e Regional, Programa de Pós-Graduação em Planejamento Urbano e Regional

Leite MP (2012) Da "metáfora da guerra" ao projeto de "pacificação": favelas e políticas de segurança pública no Rio de Janeiro. Revista Brasileira de Segurança Pública, v. 6

Leite MP (2015) Territórios de pobreza a territórios de negócios: dispositivos de gestão das favelas em contexto de "pacificação". In: Birman P, Leite M P, Machado C, Sá Carneiro S (org.). Dispositivos Urbanos e Trama dos Viventes: ordens e resistências. Rio de Janeiro: FGV

Mendes L (2017) Gentrificação turística em Lisboa: neoliberalismo, financeirização e urbanismo austeritário em tempos de pós-crise capitalista 2008–2009. Cadernos Metrópoles. São Paulo 19(39):479–512. maio/ago

Novaes PR (2018) A gentrificação periférica na Cidade do Rio de Janeiro: um estudo sobre as favelas Babilônia, Chapéu Mangueira, Vidigal e Santa Marta. Doctoral thesis. Programa de Pós-Graduação em Planejamento Urbano e Regional da Universidade Federal do Rio de Janeiro (IPPUR/UFRJ)

Ost SM (2012) O Mercado Sobe a Favela: um estudo sobre o Santa Marta pós-UPP. Dissertação (mestrado) – Universidade - Fundação Getúlio Vargas, Escola Brasileira de Administração Pública e de Empresas

Santos Junior OA (2015) Governança Empreendedorista: a modernização neoliberal. In Ribeiro L C. Rio de Janeiro: transformações na ordem urbana. Rio de Janeiro: Letra Capital, Observatório das Metrópoles

Smith N (2006) A gentrificação generalizada: de uma anomalia local à "regeneração" urbana como estratégia urbana global. In: Bidou-Zachariasen C (ed) De volta à cidade: dos processos de gentrificação às políticas de "revitalização" dos centros urbanos. Annablume, São Paulo, pp 59–87

Teixeira I, Fortunato R (2016) O turismo sobe o morro do Vidigal (Rio de Janeiro, Brasil): uma análise exploratória. Turismo e Sociedade 9(2)

Valladares L (2005) A invenção da favela. Do mito de origem à favela.com. FGV, Rio de Janeiro

Chapter 9
A New Cycle of Removals in the Favelas in Rio: What Legacy Is This?

Taiana de Castro Sobrinho

Abstract The urban restructuring carried out since 2009 in Rio de Janeiro consolidated a neoliberal entrepreneurial management model, which had been projected since the 1990s and strengthened by the accomplishment of the two largest world sport mega events in the city. Thus, Rio city is submitted to the logic of city merchandise and affected by the cultural and symbolic remeanings of certain areas of the urban space; as well it has changed the patterns of land use and occupation by deepening the relationship of the state to the private sector. These changes were possible by the establishment of a new regulatory framework that has made the urban and favela legislation more exceptional, allowing a new cycle of removals in the Rio favelas, contrary to the provisions of the Federal Constitution, the Statute of the Cities, and international treaties. Faced with this urban scenario, this article is based on the following question: How to react to such impacts of this neoliberal city project?

Keywords Favelas · Removals · Mega events

9.1 Introduction

The urban restructuring carried out in the city of Rio de Janeiro in the last 8 years consolidated a shift in the pattern of municipal urban management whose implementation had been expected since the 1990s, with the establishment of the city's first Strategic Plan[1] in 1995 with the goal, among others, to stimulate urban competitiveness in the international scenario. Such shifts were the result of a new round of mercantilization in the city in face of the boosting of the strategic planning for the

---

[1] Set of specific goals and strategic initiatives to be pursued by the municipal public power in the management of the urban territory.

T. C. Sobrinho (✉)
Observatório das Metrópoles, Institute of Urban and Regional Planing, Federal University of Rio de Janeiro, Rio de Janeiro, Brazil
e-mail: taisobrinho@hotmail.com

© The Editor(s) (if applicable) and The Author(s), under exclusive license to Springer Nature Switzerland AG 2020
L. C. de Queiroz Ribeiro and F. Bignami (eds.), *The Legacy of Mega Events*,
The Latin American Studies Book Series,
https://doi.org/10.1007/978-3-030-55053-0_9

149

two largest mega-sporting events in the world, namely, the FIFA World Cup (2014) and the Olympic Games (2016).

Thus, the adoption of the first Strategic Plan would symbolize the initial milestone of the model of neoliberal entrepreneurial management in the City of Rio de Janeiro whose goal was to submit the city to the logic of the city-merchandise, city-company, a city with an image that attracts the interests of the market and stimulates the flourishing of profitable activities (Vainer 2002).

To this end, the public administration promotes a deep restructuring both in the format of urban management and in the physical environment of the city through urban interventions that modify the patterns of land use and occupation and create new means of transport in order to facilitate mobility as well as new cultural and sporting equipment. In this context of urban entrepreneurship, these interventions take place through the intensification of the state's relationship with the private sector, having the public–private partnership as its main instrument (Harvey 2005).

In the case of Rio de Janeiro; however, there were people in the middle of such a road of destruction and reconstruction. These people lived in the favelas and saw the removal of their dwellings in a clear process of land appropriation addressed to the reproduction and economic expansion of the civil construction market, including works for building new road equipment.

Thus, our work aims to analyze the impact of this recent restructuring of the city of Rio de Janeiro on the existing housing in favelas, that is, to analyze the new removal cycle that affected the favelas of the city in the context of the mega-sporting events it held. Our focus is the removal process conducted under the urban policy practiced in the two administrations of former Mayor Eduardo Paes (2009–2016), when it became apparent that these removals were a form of state intervention in the favela territory.

With regard to the scientific justification, our goal is to contribute with a critical discussion on the issue of removals and the right to housing. As a result, our investigation is supported by theoretical frameworks of the right to the city and the critical theory of the human rights, considering a conception of human rights that is both critical and committed to concrete social practices and an alternative to the hegemonic sociolegal order marked by legal positivism and social reproduction guided by the neoliberal ideals of capitalist accumulation.

Our hypothesis was that the new removal cycle was the result of a promarket-oriented urban policy based on a strategic planning that contributed to increase the vulnerability of dwellings in favelas and to the invisibility of issues related to the protection of these dwellings. It is important to point out that no mechanisms were created to avoid the expulsion of the poor from the areas of economic interest of the city and its transfer to remote and peripheral areas, which contributed to highlight the sociospatial segregation in the urban space.

9.2   The History of the Favelas: Tolerance Versus Repression

The relationship between the public power and favelas has always been marked by ambiguity since, on one hand, the guidelines of urban legislation, from the first legal rules related to urban politics, apparently adopted an attitude of tolerance[2] in face of these rules and allowed the presence of favelas in the city landscape. On the other hand, those guidelines have always reinforced the illegal and provisional legal nature of such spaces, making clear the role played by the Law in the construction of their marginalization (Magalhães 2013).

This illegal status of the favelas ended up contributing to a precarious presence of the state in the regulation and protection of social relations in these territories, transforming them into spaces lacking infrastructure and social and urban initiatives, which further showed the dichotomy between city and favela (Gonçalves 2006) and allowed the association of favela with urban illegality.

The removals arose in this context of political construction of space (Bourdieu 1999) and of monopoly of the physical and social space determined by the unequal distribution of capital. This process ended up in tensions and disputes across the territory, according to its degree of economic valuation, and in the segregation of the dwellers who had no sufficient income to occupy the centralities.

However, according to Rolnik (2015), the logic of profitability would not only be the sole guide to these processes. They would also be guided by a discriminatory view that defines as prohibited certain forms of housing inscribed in certain sociocultural practices:

> The construction of the territorial stigma is a key element of the political gear that legitimizes expulsion. But alongside the permanent-transitory state - marking the nature of the 'reserved' land - and the territorial stigma, the hegemony construction of the registered individual private ownership over all other forms of ownership is also clearly at the origin of massive processes of dispossession (Rolnik 2015, p. 192).

Coupled with the logic of the capitalist production of urban space (Harvey 2005), what is found is the stigma of the *favelado*[3] himself, a stigma enhanced by the removal process with the goal of achieving its legitimization.

---

[2]The first occupations considered favelas in the city of Rio de Janeiro were encouraged by the public power, which prompted certain workers to occupy the slopes near their workplaces. This was the case of the Pinto Beach Favela, set up by fishermen and workers who worked on the construction of *Jóquei Clube do Brasil*, in the immediate vicinity (areas of the neighborhoods of Leblon and Lagoa, located in the South Zone, the most valued areas in the city of Rio de Janeiro), people who had been allowed to settle in the place *("Praia do Pinto acaba e deixa Ipanema que ajudou a construir"* ["Pinto beach ends and leaves Ipanema that it helped to build"], *Jornal do Brasil*, 05/11/1969).

[3]"The *favelado*, the favela's dweller, was seen as an individual that was not integrated in the neighborhood where the favela was located despite his participation in the city in many ways. At the very least, through the labour market in the area, working as construction workers, porters, cleaners, waiters etc., the *favelado* was integrated. However, it was ascribed to him the full responsibility for the existing conflicted relationship between the City and the favelas" (Brum 2011, p. 115).

Despite the emergence of new legal instruments linked to the granting of rights to favelas, including the right to nonremoval—established since redemocratization as a result of the mobilization of movements for housing—the society of Rio de Janeiro could not break with the stigma directed at the favelas and present in its imaginary, especially after the expansion of drug trafficking within their territories started in the 1990s. Such expansion perpetuated the vision of the favela as synonymous of illegal space and the main reason for the most serious problems of the city, such as urban violence and the environmental and landscape degradation of the urban environment.

9.3 A Critique of the Capitalist Production of Space

The analysis of the phenomenon of the favela removal requires a theoretical effort to contemplate its complexity beyond the legal effects, mainly focusing on its social implications. Thus, a broader approach to the meaning of the right to housing becomes necessary, and it is important to discuss it in the context of urban space and from a perspective of the right to the city (Harvey 2014; Lefebvre 2016), here conceived as one of the most important human rights—considering a conception of human rights that is critical and alternative to the social reproduction oriented by the neoliberal ideals of capitalist accumulation (Flores 2009; de Sousa Santos 2016).

Thus, in order to understand these two theories together, it is necessary to approach human rights through a reading that goes beyond formal universality and equality. Human rights should be understood not from the point of view of something given simply because they are regulated by domestic laws or international treaties and laws but rather as rights to be acquired (and built) by the exercise of social struggles and practices necessary for human reproduction and for a dignified life, that is, human rights as a dynamic process (Flores 2009; de Sousa Santos 2016).

Nonetheless, in what way would such a theoretical and practical construction be possible?

This construction will only be possible through a new culture of human rights which, in addition to breaking with abstract legal formalism, would understand Law as the interim result of conflicts and social struggles in search of emancipation and human empowerment, taking into account the historical and social context of these collective practices. Thus, Law as science and regulatory guarantees would be understood as created and recreated according to the impact of social action on reality (Flores 2009).

Considering human rights as rights that are prior and superior to any law, the conception of what a right is, in accordance with the dominant normative standard, would even include the "nonright" from an amplified and less simplistic perception than the legalist one (Lyra Filho 2005). For this reason, it is emphasized that "[...] the dialectical view must widen the focus of the Law, encompassing collective pressures (and even, as we will see, the non-state rules of oppressed and spoiled classes and groups) that emerge in civil society" (Lyra Filho 2005, p. 9–10).

It is important to note that, although certain rights are guaranteed in the established legal order, in certain territories of the city (such as the favela) and for certain social segments, respect for legal guarantees is not the rule, but rather, the exception, due to the relativization of the protection of these guarantees (Vainer 2015).

Given this, in order to understand the right to the city as a human right, this theoretical effort of the conception of Law as a break with the dominant normative standards is necessary—not in the sense of ignoring normativity, but of exceeding it. This will be achieved through a critical and realistic theoretical construction that recognizes plurality and cultural diversity, besides recognizing that alternative and counterhegemonic social practices are an essential condition for human emancipation (Flores 2009; de Sousa Santos 2016).

In this way, life in the city should be understood as a counterpoint to the sense of the city as a commodity, as a place of consumption and consumption of the place, contemplating the use-value (the city as habitat, the urban life, the encounter) to the detriment of the exchange value (the city as a commodity, the spaces bought and sold, the exacerbated consumption of goods, properties, places, and symbols) (Lefebvre 2016).

Having stated that, the question is: What would be the best way to claim that right? Through social struggles that pursue some democratic control over the process of the city configuration in search of a city that meets their collective needs and in which the sociocultural identities of each citizen are respected, as a support for the inclusion and recognition of differences and heterotopias. To these struggles, Harvey calls "insurgent citizenship," which would consist of the citizens' reaction against the city centered on the exchange value and the privatization of public spaces (Harvey 2014).

The right to the city would be consistent with the "right found on the street," which has the potential to transform public spaces addressing the action of citizens, that is, the subject, as a way to ensure such direct democratic participation (Sousa Júnior 2016). After all, this right emerges from the voices of the streets, encounters, and conflicts around which the city will be built as a result of social mobilization and political struggle to reconfigure urban space. It will set up an image that goes beyond a city shaped by real estate capital, public–private partnerships, and state action committed to urban entrepreneurship, which are currently the great owners of control over the surpluses derived from urbanization.

9.4   Strategic Planning and Urban Entrepreneurship

The centrality of the strategic plans in the urban planning of the last years in the City of Rio de Janeiro[4] consolidated a model of neoliberal entrepreneurial management, in which the public administration submits the city to the same logic in which the

---

[4] Although the first Strategic Plan was established in 1995, it only gained centrality in urban politics from the first administration of former Mayor Eduardo Paes (2009–2012). After this, it became

companies work: competition for attracting financial and real estate investments. According to Carlos Vainer, "[...] this management project implies the direct and immediate appropriation of the city by globalized business interests and depends to a great extent on the banning of politics and the elimination of conflict and the conditions for exercising citizenship [...]" (Vainer 2002, p. 78).

With this, the city acquires the status of subject, personified through the features that allow advantages in the competitive dispute with other cities of the globalized world, a city that must be sold because of its qualities in order to arouse the interest of the international financial capital, that is, the *"Rio Cidade Maravilhosa"* [Rio Wonderful City], a tourist city with stunning natural landscapes.

Urban marketing is a basic tool of this type of management. What is sold are the landscapes, locations, lifestyles, regional identities, that is, the features that make the city unique and distinct from all others—the symbolic power of the collective capital of the city (Bourdieu 1989). The image of prosperity and innovation is what is sold, and the social difficulties that the city may face are hidden, that is, "[...] the triumph of the image over substance is total [...]" and any manifestation of poverty or inequality must be made invisible (Harvey 2005).

The onset of such transformations was both the creation of the 2009–2012 Strategic Plan, implemented with the election of Eduardo Paes to the City Hall of Rio de Janeiro in 2009 and the General Law of the World Cup.[5] These two legal instruments would enhance a pattern of entrepreneurial management in the urban politics and legislation since they would relativize the existing legal order through decrees and new legal frameworks.

Thus, there is no way to disassociate urban remodeling in the city in the last 10 years from the mega events that the city hosted during this period. Events of such size offer new dynamics of capital circulation in the city's environment through public–private partnerships (key instrument of this renewal process), expansion of the real estate market, and of the service sector, besides the incentives resulting from cultural events.

In this sense, taking into account that the cities hosting these events have features that are common to them, the Thematic Report on Mega-sporting Events (2009) of the Human Rights Council of the United Nations for Adequate Housing highlighted the denunciations of forced removals and evictions, in addition to denunciations "[...] of reduced access to housing as a result of gentrification, of large-scale operations against the homeless, and punishment and discrimination against marginalized groups"[6] (A/HRC/13/20, 2009/Rolnik 2010, p. 3).

---

mandatory and must be submitted up to 180 days after the inauguration of the mayor from the Amendment to Organic Law N° 22 of 2011, which added article 107-A to it.

[5]Law N° 12663 of June 5, 2012.

[6]Removals, or forced evictions, are phenomena that portray the intersectionality of oppressions as they are evidence of discrimination regarding a social and geographical class, that is, poverty and the favela, thus covering other discriminatory relationships such as gender and race. In this context, it should be noted that "[...] urban spaces, created by local acts of elimination of favelas and forced segregation, are the spatially marked manifestations of the marginalization of race, gender and class [...]" (Perry 2012, p. 170).

In addition, such entrepreneurial logistic management of a city opposes the guidelines of the *Estatuto da Cidade* [City Statute][7] (Law N° 10257/2001), contrary to the principle of democratic management of urban policy (articles 43–45) and to a city committed to the use value and to a social function of property and city. In addition, it also is contrary to the provision set forth at the statute that called for an urban policy ruled by the Master Plan,[8] the main legal instrument of the municipal administration since it promotes and sustains class alliances between the public power and the private sector which revolve around this city-business concept (Castro and Novaes 2015).

It is worth noting that the logic of the capitalist production of space does not block any development guided by initiatives other than those of urban entrepreneurship. Even in a context of expansion of promarket-oriented management, we can find social practices that are resistant and insurgent to this model of a city, establishing a counterpoint to the exchange value given to housing as well as to social and cultural manifestations of the appropriation of urban space. What we attempted to argue here is that the financial capital has the mean to turn almost everything into commodity, including local culture and lifestyles that symbolize the identity of a given city, appropriating and extracting surpluses from the distinctive landmarks of each city.

9.4.1 Mega Events as Opportunity

From the 1970s onward, the association of the sporting events of international projection with the processes of urban restructuring became more evident. According to Rolnik (2010), this scenario would be even more significant in the 1990s:

> In the 1990s, the practice of organization of mega events as elements of a strategic urban planning with the goal of improving the position of the chosen cities in the globalized economy became hegemonic. The realization of international games as a strategy for economic development, including the renewal of urban infrastructure and real estate investment, has become the contemporary approach to mega events by cities and States (Rolnik 2010, p. 2).

The discourse surrounding major events, notably mega-sporting events, allows the mobilization of the population around a city project and its consequent urban interventions as a way of preparing for these occasions, triggering the ideas of legacy and common good. Thus, a certain social consensus is generated in relation to the need of urban interventions for the restructuring of the city by means of the implementation of sports, road equipment, and other infrastructures because the image that is sold is that "who wins is the population.[9]".

---

[7] National law that establishes the general directives for urban policy and sets forth other provisions.

[8] Plan approved by Municipal Law, and mandatory for cities with more than 20 thousand inhabitants. It consists of the basic instrument of development and urban expansion policy and is part of municipal planning.

[9] Available at <http://ultimosegundo.ig.com.br/brasil/rj/olimpiadas+sao+used+as+pretexto+for+investments+no+rio/n1237964716097.html. Accessed: May 3, 2017.

In this way, it is very important to investigate the territorial, socioeconomic, and symbolic transformations promoted by the mega events held in the City of Rio de Janeiro—not only the 2014 World Cup and the 2016 Olympics but also the 2007 Pan American Games which started a long period of the city as the seat of several mega events—in addition to the impacts of such transformations on the right to housing and the right to the city.

The relationship between major urban restructuring projects and massive processes of expulsion of the poor population from their dwellings is also clear when we analyze the specific goals of the city's last Strategic Plan (2013–2016) created in the context of the removal cycle, that is the object of the current work. Its specific goal is to reach "[…] at least 5% reduction of areas occupied by favelas in the city until 2016, base year 2008 […]" for the housing and urbanization sector (Prefeitura do Rio de Janeiro 2013).

Another specific goal was to "[…] complete, until the end of 2016, the redevelopment works of the Porto Maravilha project […]" (Strategic Plan/Prefeitura do Rio de Janeiro 2013). At this point, it is important to clarify that this project established an *Operação Urbana Consorciada* (OUC) [Urban Partnership Operation][10] with the goal of performing the urban renewal of the port area of the city of Rio de Janeiro, a territory previously abandoned and devalued by the real estate market.[11]

Thus, it is possible to conclude that initiatives involving the strategic planning of Mayor Eduardo Paes' administration (2009–2016) deepened and sharpened the proposal of a neoliberal city. In this period, the urban planning of the city was negotiated, and an entire area was privatized—as is the case of the port area—through strategic goals and initiatives to be conducted through public–private partnerships. As a result, a new form of relationship between the state, private sector, and city was consolidated.

Many and diverse were the impacts on urban space derived from this logic of market-friendly planning (Vainer 2015). However, their reflection on the right to housing of those who lived in urban occupancies and favelas in Rio de Janeiro consisted of the forced removal of approximately 77,206 people from their dwellings (Comitê Popular 2015). It also consisted in the expansion of both the exchange value and real estate speculation even for housing in favelas, changing the dynamics of the

---

[10]The Urban Partnership Operation (OUC) consists of a set of interventions and measures coordinated by the municipal public power, with the participation of owners, residents, permanent users, and private investors, with the objective of achieving structural urban transformations, social improvements, and environmental valuation (article 32, of Law N° 10257, 2001). The OUC of the port area of Rio de Janeiro—the largest in the country—considered a Special Urban Interest Area, was created by Complementary Law N° 101 of November 23, 2009, with the purpose of urban redevelopment of an area of 5 million m^2, under the management and supervision of the *Companhia de Desenvolvimento Urbano da Região do Porto do Rio de Janeiro* (CDURP) [Rio de Janeiro Port Region Urban Development Company], encompassing the neighborhoods of Santo Cristo, Gamboa, Saúde and sections of Dowtown, Caju, Cidade Nova and São Cristóvão. For further information about this OUC, see http://www.portomaravilha.com.br/portomaravilha.

[11]An area of the memory and culture of the black population.

social relations that until then were developed there,[12] and consequently increasing the peripheralization of the poor population besides intensifying the segregated urban spaces.[13]

9.5   The Grammar of Removals: Discourses and Justifications

According to what has been seen and by analyzing the historical relationship between removal and dwellings in favelas, it is possible to perceive that removals are used as a measure of social and urban space control, mainly of the expansion control of the growth of occupancies and spaces in the favelas (particularly in areas attractive to the real estate market).

On some occasions, forced removals were conducted in an arbitrary and violent way, involving threats, psychological terror, and even physical assaults on the dwellers. In addition, many eviction episodes occurred through nocturnal incursions and with tight deadlines for vacancy. In this way, the deadlines urgency for the onset of restructuring was used as a reason for the violation of the rights of the affected population, especially the right to participation and information on these urban development projects (A/HRC/13/20, 2009/Rolnik 2010).

In addition to the forced removals, another form of displacement may also be identified as an indirect consequence of these urban projects to improve the city's image, such as the expulsion by the market due to the increase in the cost of housing. In these situations, it is necessary to leave the house where one resides because of the increase in the price of the property due to the lack of financial conditions to stay on the site, particularly when it comes to rent.

In this way, the area affected by the 'white removal' underwent a deep alteration in its demographic composition and in the social dynamics that had been developed there until then. This occurred since the previous inhabitants were eventually pushed to peripheral areas of the city and, as a result, lost their community ties and lost their jobs most of the time. In addition, they had to face the impact of increased transportation costs to reach their workplaces, places close to the location in which they had lived.

Besides the legacy's discourse, this new cycle of removals in the city's favelas was also made possible by a legal framework that has created exceptions in both

---

[12]Such economic valuation was noted, for example, in the Vidigal favela, located on the South Zone of Rio, next to the fashionable neighborhood of Leblon, one of the most expensive regions of the city. In the period of 2008–2014, Leblon showed an increase in 477.3%, reaching the amount of R$ 8,370.00 per m^2 in 2014 (Castro and Novaes 2015).

[13]More than half of the families removed were resettled in low-income housing developments in peripheral areas of the city in need of infrastructure and public services, according to the Report titled *"Entenda a Política Habitacional da Prefeitura do Rio"* ["Understanding the Housing Policy of the City Hall of Rio"] (2015).

the urban and favela legislation, making the principle of nonremoval[14] more flexible in a way that their dwellers were deprived from their rights, including those who lived in consolidated and entitled favelas, with years of existence.[15] In addition, most of these interventions were ruled by Municipal Decrees, that is, instruments of exclusive competence of the Chief Executive, the mayor, instruments which do not require legislative discussion and approval, as a law would.

Finally, it is worth noting the influence of popular mobilizations and resistances on the processes of city restructuring. Sometimes such influence can interfere in the negotiations involving the affected people and the public sector and incorporate certain demands of the local population. In this way, the presence of conflict in urban planning decisions is very important and should be highlighted since, without social and political tensions in the implementation of these projects, they would be envisaged as antidemocratic, arbitrary, contrary to the overall urban legislation, and to the Constitution itself which provides for the democratic management of urban policy.

With regard to the new legal instruments created in this context to support and facilitate the implementation of this type of intervention in favela areas, highlights are the 2011 Master Plan[16] (after the City Statute), the expropriation decrees, and the decree that regulates the urbanistic freezing of favelas. They are based on public interest and are issued by the City Hall of Rio de Janeiro in recent removal processes.[17]

Such legislative construction opposes the urban practices that had already been set up by the dwellers in these spaces, not only delegitimizing the existing formats of housing in the favelas—which are part of the history of the urban development of

---

[14] Article 429, item VI, of the Organic Law of the Municipality of Rio de Janeiro establishes that: "Art. 429—Urban development policy shall respect the following premises: (…) VI—urbanization, plot regularization, and entitlement of the favela and low-income areas, without eviction of the residents, except when the physical conditions of the area occupied are life threatening to the residents, a hypothetical situation in which the subsequent rules will be followed: a. technical evaluation of the responsible institution; b. participation of the interested community and representative entities on the analysis and definition of solutions; c. relocation to areas near the housing or working sites, if displacement is necessary". Available at: https://leismunicipais.com.br/lei-organica-rio-de-janeir o-rj. Accessed: July 19, 2016.

[15] Like the *Vila Autódromo* favela, located in the neighborhood of Jacarepaguá (real estate expansion area), which despite actually having a concession of use for housing purposes for a period of 99 years (granted in the 1980s) had approximately 530 families removed from the 550 families who used to live there at its onset.

[16] In addition to not providing for the nonremoval principle—removal was a word that was simply excluded from its legal text, and replaced by the word relocation, according to the wording of article 15, paragraph 2, of Complementary Law N° 111 of 2011 and of article 3, sections IV, V, and VI, which provides for the restraint of land and urban development irregularities; the urbanization of favelas and irregular subdivisions, aiming at their integration into the formal areas of the city, with the exception of situations of risk and environmental protection; and the restraint of the growth and expansion of favelas, through the establishment of physical limits and special urban rules, respectively.

[17] Decree N° 33648, of April 11, 2011. This context is accentuated because this decree determines the urban freezing in the favelas as it revokes the right to construct new buildings in favelas in the *Área de Especial Interesse Social* (AEIS) [Area of Special Social Interest].

the City of Rio de Janeiro—but also making difficult their process of regularization, since what is attempted is to apply reality to the model and not the model to reality in a consistent way, a way able to understand the differences and peculiarities of each favela.

The step-by-step description of the removal process is not an easy task since all the information regarding this process is difficult to access, notably data of the favelas targeted to be removed as well as the reasons for the removal and also the urbanization projects used as justification for this action.

In addition, updated data on the number of the removed population and the provisions offered to these families as a form of compensation are not easily disclosed by the secretariats attached to the municipality, which are the bodies responsible for storing this information. As these data are often considered classified, it is worth pointing out that the country has a system that provides for transparency in public management and removal is a measure that impacts on the collective and fundamental rights of all those affected.

The particularities of the removal can be known from reports of the social actors involved in this process, such as the affected dwellers, the reports, and the action of public defenders, the leaders of social movements, as well as from documentaries focusing on the way the removals were conducted. In this way, it is possible to detect a pattern in the state's action within the favelas as far as removals are concerned, with the following variations: (i) size of the area to be removed, (ii) reason given for the removal, (iii) negotiated compensatory measures, (iv) intensity of the use of violence, and (v) degree of the resistance found.

A characteristic feature of the last removal cycle experienced in the City of Rio de Janeiro was the complete absence of a collective negotiation since the initial approach, performed by City Hall agents, was generally carried out individually and through proposals of compensatory measures that differed from one family to another. Furthermore, as the negotiation process was always carried out by word of mouth, no documentation on the process was available, and sometimes not even produced.

Thus, when the removals are analyzed from a spatialized perspective, the conclusion is that these eviction processes, in general, served not only as vectors for the intensification of the sociospatial segregation in the City of Rio de Janeiro but also for the economic valuation of the areas of expansion and centrality. In this sense, as the urbanist Ermínia Maricato highlighted, the strategy was to "clean up the city" through the economic renewal of these territories (Maricato 2015).

9.6 Conclusion

To conclude, the positive legacy, understood in this work as a result of the overall process of spoliation of housing and of other human rights—apart from any physical legacy that the physical interventions of improvement in the city can represent—was the permanence of a few favelas that had their complete removal announced.[18] The dwellers who resisted the removal and claimed their right to urban housing and their right to the city—here understood as the right to act directly in the urban processes and to recreate and propose the city where one lives—symbolize the conquest of emancipation and awareness as subjects of rights, the metamorphosis proposed by the critical theory of Law for the oppressed and dispossessed.

The self-recognition as citizens and subjects of rights opened spaces of struggle and contestation to the existing logic of the city, allowing the emergence of new perspectives that will transform the reality of the city, so much homogenized and linked to the promotion of Rio de Janeiro in global city circuits, chiefly highlighted after the mega events held in the city in recent years.

Thus, the insurgency of the favela dwellers regarding the public initiatives to relativize, and even to withdraw their rights, transforms the favelas into political and alternative spaces for the creation of Law and of a right understood as a manifestation of the principles of a legitimate social organization of freedom, the right as freedom, the very "right found on the street" (Souza Júnior 2011). A right that results from the "subaltern cosmopolitanism," which consists of the emancipatory practices for claiming social inclusion and for overcoming the conditions of oppression and plundering of rights (de Sousa Santos 2016).

References

Bourdieu P (1989) O poder simbólico. Bertrand Brasil, Rio de Janeiro
Bourdieu P (1999) La miséria del mundo. Akal Editorial, España
Brum M (2011) Cidade Alta: história, memórias e estigma de favela num conjunto habitacional do Rio de Janeiro. Ponteio Edições, Rio de Janeiro
Castro DG, Novaes PR (2015) Empreendedorismo urbano no contexto dos megaeventos esportivos: impactos no direito à moradia na cidade do Rio de Janeiro. In: Castro DG et al (ed) Rio de Janeiro: os impactos da Copa do Mundo 2014 e das Olímpiadas 2016, 1st edn. Letra Capital, Rio de Janeiro
Comitê Popular da Copa e das Olimpíadas do Rio de Janeiro (2015) Megaeventos e violações de direitos humanos no Rio de Janeiro. Rio de Janeiro, Comitê Popular da Copa e Olímpiadas do Rio de Janeiro. https://issuu.com/mantelli/docs/dossiecomiterio2015_issuu_0. Accessed 18 Feb 2016
de Sousa Santos, B (2016) As bifurcações da ordem: revolução, cidade, campo e indignação. Cortez, São Paulo
Flores JH (2009) A (Re) invenção dos direitos humanos. Fundação Boiteaux, Florianópolis
Gonçalves RS (2006) A política, o Direito e as favelas do Rio de Janeiro: um breve olhar histórico. Urbana: Revista Eletrônica do Centro Interdisciplinar de Estudos da Cidade (CIEC), Universidade Estadual de Campinas, Departamento de História & Instituto de Filosofia e Ciências Humanas, ano 1, n. 1, set./dez
Gustin M, Dias MT (2006) (re) Pensando a pesquisa jurídica: teoria e prática, 3rd edn. Del Rey, Belo Horizonte
Harvey D (2005) A produção capitalista do espaço. Annablume, São Paulo

---

[18]Like the symbolic cases of the *Estradinha* (neighborhood of Botafogo) and *Vila Autódromo* (neighborhood of Jacarepaguá) favelas located in areas of intense real estate valuation and incidence of conflicts as a result of their appropriation.

Harvey D (2014) *Cidades rebeldes:* do direito à cidade à revolução urbana. Martins Fontes, São Paulo

Lefebvre H (2016) O direito à cidade. Centauro, São Paulo

Lyra Filho R (2005) *O que é o direito.* Livraria Brasiliense, São Paulo

Magalhães AF (2013) O direito das favelas. Letra Capital; FAPERJ, Rio de Janeiro

Maricato E (2015) Para entender a crise urbana. Expressão Popular, São Paulo

Perry KKY (2012) Espaço urbano e memória coletiva: o conhecimento de mulheres negras em lutas políticas. In: Santos RE (ed) Questões urbanas e racismo. Ed. De Petrus and Alii Editora, Petrópolis (RJ)

Prefeitura do Rio de Janeiro (2013) *Plano Estratégico da Cidade do Rio de Janeiro Pós-2016:* o Rio mais integrado e competitivo

Rolnik R (2010) Relatório temático sobre megaeventos esportivos. A/HRC/13/20, 2009. Escrito com a colaboração de BrendaVukovic. A tradução para o português, utilizada neste trabalho, foi realizada pela ONG FASE, em novembro de 2010

Rolnik R (2015) *Guerra dos lugares:* a colonização da terra e da moradia na era das finanças, 1st edn. Boitempo, São Paulo

Sobrinho, TC (2017) *O novo ciclo de remoções nas favelas do Rio de Janeiro:* que legado é esse? Master's thesis. Programa de Pós-Graduação em Direito. Faculdade Nacional de Direito da Universidade Federal do Rio de Janeiro (FND/UFRJ). Orientation of the master's thesis: Mauro Osória da Silva. Alex Magalhães, Co-orientation

Souza Júnior JG (2011) *Direito como liberdade:* o Direito achado na rua. Sergio Antonio Fabris, Porto Alegre

Vainer C (2002) *A cidade do pensamento único:* desmanchando consensos—Otília Arantes, Carlos Vainer, Ermínia Maricato. Vozes, Petrópolis (RJ)

Vainer C (2015) Cidade de exceção: reflexões a partir do Rio de Janeiro. Anais do XIV Encontro Nacional da ANPUR, Rio de Janeiro, 2011. http://br.boell.org/site/defaut/files/downloads/carlos_vainer_ippur_cidade_de_exceção_reflexões_a_partir_do_rio_de_janeiro.pdf. Accessed 6 Jan 2015

Chapter 10
State of Art and Possibilities for Citizenship Education in the City of Rio de Janeiro

Filippo Bignami and Ana Paula Soares Carvalho

Abstract This chapter scrutinizes what is the meaning of educate to citizenship and tries to scale this concept in the case of the educational system in Rio de Janeiro within the project "Urban regime and citizenship. A case study for an innovative approach." Citizenship education is, from many years, the center of a lot of public debates in many western countries and it is also subject of lively theoretical debate that involves many researchers afferent to different disciplines. There are many different reasons that justify the actual interest in citizenship education in democratic political context. We discuss the efficiency of citizenship education, on its effective capacity to contribute in social cohesion with the education of a citizen who is able to cooperate with whom thinks differently and follows a different lifestyle. Henceforth, we debate the state of citizenship education in Rio de Janeiro's public secondary schools and how improvements in this area could be fruitful in this context. We then present our idea for a module for citizenship education to be used as a learning tool by teachers of public schools.

Keywords Citizenship · Education · Urban setting

10.1 Introduction

The concept of citizenship is nowadays debated and under tension, often inappropriately called "a determinative legal status" (Nguyen 2018, p. 93), simply something

F. Bignami (✉)
Department of Economics, Health and Social Sciences, DEASS - Labour, Urbanscape and CItizenship, LUCI Research area, University of Applied Sciences and Arts of Southern Switzerland, SUPSI, Lugano, Switzerland
e-mail: filippo.bignami@supsi.ch

A. P. Soares Carvalho
Department of Social Sciences, Pontifícia Universidade Católica do Rio de Janeiro (PUC-Rio), Rio de Janeiro, Brazil
e-mail: apcarvalho@gmail.com

L. C. de Queiroz Ribeiro and F. Bignami (eds.), *The Legacy of Mega Events*, The Latin American Studies Book Series, https://doi.org/10.1007/978-3-030-55053-0_10

of an analytical–reconstructive path. It rather goes beyond that, however, as a nexus between individuals and society, as it identifies the political, social, economic, and cultural characteristics of such a nexus. It is, in fact, identified in specific dimensions, affecting different theoretical aspects (stemming from both political and social studies) and practical aspects (the legal and administrative). Concerning citizenship, a pivotal role is played by the individual as the main actor of this nexus.

The individual is the key driver and needs to acquire knowledge of how to "join" the society as its member. Acquiring the capacity to be a politically, socially, culturally, and economically active member of the society is therefore a fundamental component of citizenship education. A distinction should be made between the behavior of citizenship and the components of the competences on which this behavior is built. The components of competences are formulated in terms of knowledge, attitudes, skills, and reflection (Geboers et al. 2013).

Much has been discussed of the positive aspects of citizenship education programs in primary and secondary schools. It is widely held that citizenship education should focus on empowering students to assume an active role in the process of defining and expanding citizenship itself (Menezes 2003; Isin 2017a; Lin et al. 2015). Further, it enables youth to "learn by doing," to put their political and social skills and knowledge into action (McIntosh and Youniss 2010). This would promote student orientation toward acting on contingent citizenship. Other scholars, linking the decline of actively choosing capacity, in turn, addresses the importance of developing citizenship learning situations to engage students as "active change agents" instead of clients or consumers (Barber 2008; Pinkett 2000). Or, as Warleigh would put it, citizenship education should move from the "knowledge about" to "action," mobilizing students' experiences (Warleigh 2006).

This chapter aims to describe the experience of implementing a pilot citizenship education module that entails the usage of active teaching methodologies—in this case, debate (Kennedy 2009, 2007), particularly choosing the speed format, and PBL (Problem Based Learning) (Panlumlers and Wannapiroon 2015)—in a secondary public school in the city of Rio de Janeiro. We believe that sharing this experience can be useful for those interested in elaborating strategies of dealing with the subject of citizenship in school environments not only, but especially, in cities with a complex social and political texture, facing specific challenges in terms of impairments to significant progress in democratic conditions.

In this chapter, we firstly present a theoretical discussion on the potential contributions of citizenship education for the development of students' sense of political and social membership. In the second section, we briefly discuss citizenship education in Brazil, in the city of Rio de Janeiro, as well as its implications in such a context. The third section describes how we have elaborated the citizenship education module and its outlines and briefly present the module test in one school. We conclude the chapter with a short discussion of the results, highlighting the need to increase the effort in citizenship education.

10.2   How Citizenship Education Can Contribute to the Development of the Sense of Political and Social Membership in Future Citizens

Over recent decades, citizenship education has become a key issue in numerous academic and public debates. It is also the subject of a lively ongoing theoretical debate, which involves researchers from different disciplines. The criteria that characterized the definition of citizenship, from its most influential version in the period after the Second World War (Marshall 1950) until the end of the last century, are no longer sufficient. New meanings of the citizenship concept are visible in a modified social and political context: the aggressive intrusion of information and communication technologies in the public field, as well as in private life and in the labor market, the economic, and financial interdependence triggered by so-called "globalization," the increasing societal inequalities in wealth distribution together with youth unemployment and migration flows (WEF 2018), have influenced and contributed to redrawing the idea of citizenship in contemporary society.

Furthermore, migration nowadays is a key variable impacting the concept of citizenship, evidencing manifold forms of participation, identity, ways of exploiting rights and duties, and membership. Such impact is effectively traced and gauged in the process of "differential inclusion" (Mezzadra and Neilson 2012, p. 67) framed by Mezzadra and Neilson. Such slant of the concept of citizenship allows, on the one hand, "the emergence of locations of citizenship outside the confines of the national state" (Sassen 2002, p. 281) and an increase, on other hand, in the role of cities in shaping a modern conception of urban citizenship (Alsayyad and Roy 2006). These enrichments of the concept of citizenship contribute to understanding why citizenship education is actually a seminal, though debated and questioned, issue, involving the question of how to make citizens aware of their individual potentialities, roles and responsibilities, connecting, in other words, city and citizens with the concept of citizenship.

There are basically two important reasons to justify the present interest in citizenship education from a political perspective.

First, for a civic reason: there is an increasing need to identify citizenship models suited to guide the actions of public education institutions. The solutions may vary, depending on the criteria chosen to lay out the values on which social, political, and civil common shared life should be based in democratic societies, and from the context: e.g., citizenship education must be set differently in Europe and in South America. In this context, the concept of social cohesion has become of major interest. In fact, issues regarding the "cement of society" (Elster 1989) have been present in numerous important public and academic debates in past years, such as the one on repositioning citizenship in a frame of social cohesion maintenance beyond the nation–state; or rather, in a persisting formal frame of nation–states, the destabilization of polities as national state-centered hierarchies of legitimate power and allegiance has enabled a multiplication of nonformalized or only partly formalized political dynamics and actors (Sassen 2005).

These dynamics enable the deterritorialization of citizenship practices in terms of participation, membership, and identities, and of discourses surrounding allegiance and loyalty. Further, in context with a development of unequal conditions of citizenship, like in Rio de Janeiro, where the mechanisms of reproducing social inequalities are manifold in the schools (Koslinski and de Queiroz Ribeiro 2017, p. 167), such education is a means to increase individual mobilization and responsibility to participate in order to claim social and political rights and roles. Education towards citizenship from a civic perspective is the main way to "invest" in all the main types of learning outcomes (UNESCO 2015), namely: cognitive (knowledge, understanding, and critical thinking); socioemotional (sense of belonging and sharing responsibilities); behavioral (acting effectively and responsibly, motivation to take action).

Some authors use the adjective "insurgent" to depict the citizenship emerging from Brazil's urban settings, in particular, peripheries (Holston 2008). Holston points out how it is characterized by inegalitarian citizenship with inclusive membership, since the politics of properties and land show how most Brazilians have historically been excluded from the legal use of land, forcing them to live illegally and making illegality an important condition of settlement. He is aware, then, that the expansion of membership and identity as dimensions of citizenship has expanded along with "*new kinds of violence, injustice, corruption, and impunity*" (Holston 2008, p. 13).

The second reason for an increased interest in citizenship education from a political perspective: presently pluralism is often seen from a "cultural" dimension, questioning in what terms citizens are incorporated into a polity through a cultural driver. An effective definition of a citizen is: "*(...) a member of a political community, entitled to whatever prerogatives and encumbered with whatever responsibilities are attached to membership.*" (Walzer 1989, p. 211). This implies that at least an efficient citizenship education needs to lean on a theory providing an account not only of the rights and civic obligations of citizens, but also of who citizens are, the context and conditions in which they need to interact, both in proximity and in wider arenas, and on what basis they are to be "incorporated" into the political community, beyond the cultural aspect. Citizenship, then, is supposed to contribute to the strong development of a sense of political and social membership in citizens. This involves political and social "performative citizenship" (Isin 2017b) by a revival over: *who may and may not act as a subject of rights; (...) social groups making rights claims; people enact citizenship by exercising, claiming, and performing rights and duties; when people enact citizenship they creatively transform its meanings and functions* (ivi, p. 501). In turn, we question the efficiency of citizenship education in terms of its capacity to promote social cohesion by educating citizens able to cooperate in a performative sense with fellow human beings who think differently and who follow different lifestyles.

A fundamental question that the theory of citizenship education actually has to answer is: which citizen? Which citizenship? Different answers are possible. In fact there exist diverse models of citizenship that propose various solutions to the challenge of promoting social cohesion in a context marked by pluralism (Westheimer and Kahne 2004). From this point of view, important citizenship education concepts include civic minimalism by Galston (1991, 2005); political liberalism by John Rawls (Rawls 2005), and the different versions of republicanism, from the classical republicanism raised by Maynor (2003) to the liberal one by Dagger (1997). Each of these concepts involves opportunities and risks that each perspective of citizenship education takes into consideration; a fruitful balance should be found in defining contents fitting each specific context, in a frame in which the positive effects of citizenship education in the secondary school (both in the curriculum in school, out of school, and extracurricular), in the classroom and on students' awareness of the meaning of being citizens, are demonstrated from various research (e.g., Geboers et al. 2013).

In this sense, democratic conditions and social vulnerability are tackled since the objective is to foster citizenship awareness and can then be envisaged mainly (but not only) in two ways: a) as a remedy to the risk of exclusion of students, especially those coming from disadvantaged contexts and b) "*Citizenship is not something given, but must be fought for and claimed*" (Bignami et al. 2016, p. 34) and is mandatory to encourage signals and patterns of participatory citizenship dynamics in sociopolitical actors not conscious of having such potentiality (citizens not aware of being key players for citizenship).

A citizenship education process needs to develop democratic conditions and contributes to the prevention of social vulnerability and the segregation effects revealed in Rio de Janeiro (Koslinski and de Queiroz Ribeiro 2014), promoting the development of participatory citizenship dynamics. In these conditions, students can challenge the "spatial" division between the private and public spheres, where a conceptual barrier is usually placed, hampering people to enable a horizontal (link from individual to individual and from individual to collectivity) and a vertical (individual-participatory instances/political system-state) citizenship. Citizenship education appears, therefore, as a powerful means to reduce inequalities and increase social mobility (Corak 2016).

10.3 Citizenship Education in Brazil and in the City of Rio de Janeiro: A Brief Overview

Until the secondary school reform of 2017, secondary schools in Brazil were divided into regular and technical secondary schools. Regular secondary schools are 3 years and attended by students aged 15–17 and technical secondary schools are usually 4 years long. In both cases, there is a basic required curriculum in all schools. The mandatory subjects are: Portuguese, mathematics, physics, chemistry, biology, philosophy, sociology, history, geography, and arts. Philosophy and sociology became mandatory in 2009.

In Brazil, citizenship education is not offered as a secondary school subject or a specific course. The topic of citizenship is spread across different subjects, especially the humanities (philosophy, sociology, history, and geography).

Most of the secondary education in Brazil is funded and organized by the states—Brazil is a federative republic, so states have some autonomy and some specific obligations in terms of offering basic public services. Despite this level of autonomy, they are required to observe some federal regulations, including school curriculum.

The Law of Directives and Bases of National Education (LDBE), enacted in 1996, defines that the official curriculum is based on parameters established by the Ministry of Education, in conjunction with sectors of the civil society, organized around the National Education Council. Until 2017, private and public secondary schools were required to follow the National Curricular Parameters (PCNs) and the National Curricular Guidelines (OCNs).

Although the topic of citizenship appears in all these official documents, there is not a specific governmental initiative to prepare teachers to deal with this subject in the classroom, especially concerning not only cognitive, but also socioemotional and behavioral aspects (UNESCO). Mechanisms for learning assessment in the realm of citizenship education are also virtually inexistent. It is worth mentioning, however, that all undergraduate teacher training programs must include a course load dedicated to citizenship and humanrights[1], but they do not necessarily include preparation for the teaching of these subjects.

In the State of Rio de Janeiro, the state public schools follow curricular guidelines defined by the Secretary of Education. Following the PCNs and the OCNs, the state defined a minimum curriculum, including the defined skills and competences, students are supposed to develop. The curriculum is divided into 2-month period, each with a determined theme and set of skills and competences.

As could be expected, the topic of citizenship appears sparsely in the realm of subjects connected to the humanities and there are no specific programs of teacher preparation focused on citizenship education.

In the minimum curriculum, the word "citizenship" usually appears related to topics such as human rights, civil society, state, constitution, politics, and participation. Those skills and competences cover a good amount of the formal aspects of citizenship, but focus mostly on cognitive aspects and very little on the socioemotional or behavioral. The acquisition of knowledge in comparison to other abilities is prevalent. For example, in the minimum curriculum sociology requirements for the second and third year, the language adopted is oriented to knowledge (e.g., "Understand the concept of citizenship and the historical emergence of civil, political, social and cultural rights as a continuous and expanding process; Understand the historic role of social movements in the construction of citizenship[2]", etc.). The verb "understand" dominates, whereas expressions such as "experience," "develop attitudes," and "enact," among others related to the socioemotional and behavioral fields, are almost absent.

---

[1] See http://portal.mec.gov.br/index.php?option=com_docman&view=download&alias=70431-res-cne-cp-002-03072015-pdf&category_slug=agosto-2017-pdf&Itemid=30192.

[2] Governo do Rio de Janeiro. Currículo Mínimo—Ensino Fundamental Anos Finais e Médio Regular. Available at: http://www.rj.gov.br/web/seeduc/exibeconteudo?article-id=5776111.

Teacher trainees not only receive no specific training to deal with citizenship-related subjects, but they also suffer from a general lack of proper working conditions in the state of Rio de Janeiro. They have too many students in their classes, sometimes over 40, usually work in more than one school, and are underpaid. In many cases, the subjects they teach are unrelated to their major fields of study.[3]

Only more comprehensive research involving the collection of data both from students and teachers from the whole city of Rio de Janeiro would allow scholars to make more accurate statements about the state of citizenship education in Rio de Janeiro, but that was not possible in the scope of this particular project. A question-naire[4] was elaborated, though, to assess students' basic knowledge of the political system in Brazil, their relationship with media and social media, their perception of the meaning of citizenship, their level of interest in public issues, their involvement with and interest in student politics, and how much they think they are learning in school about topics related to citizenship. The collected information offers some insight.

From what was gathered in terms of knowledge of and interest in formal politics, a high level of distrust and disinterest is inferred. Although most students knew of the main Brazilian political parties, around 15% of them were not able to name any of them.[5] The majority of them do not correlate formal party affiliation with being a good citizen. Even voting in every election is not deemed as an important part of being a good citizen.[6]

While most of them reported knowing the most important newspapers of south-eastern Brazil, almost 20% answered that following the news about politics on TV, newspapers, and/or the internet was of little importance when it comes to being a good citizen. As for social media, they use it mainly for communicating with friends and not to keep up with politics.[7]

Most of them take part in some kind of community work, but a much smaller part participated in groups that discuss subjects such as racism, gender, human rights, politics, environment, and animal rights.

---

[3] 27.3% of teachers of the state schools of the city of Rio de Janeiro do not have a teaching degree adequate to the subject they teach (Indicadores Educacionais 2017, INEP). Available at: http://por tal.inep.gov.br/indicadores-educacionais.

[4] The questionnaire can be accessed here: http://www.supsi.ch/deass/ricerca/banca-dati-progetti/in-evidenza/Urban. It was applied to 152 secondary students (138 second-grade students and 14 third-grade students) of the school in which the citizenship education module was tested.

[5] The question is "Cite Brazilian main political parties." It is an open question.

[6] The question is "In your opinion, in order to be a good citizen, how important are the following behaviors?" The alternatives are: to participate in discussions about social, political, and/or envi-ronmental subjects; to participate in peaceful demonstrations against government policies deemed unfair; to take action in order to protect the environment; to vote in every election; to be affiliated to a political party; to follow the news about politics on TV, newspapers, and/or the internet. A Likert scale was used in this question (very important, important, of little importance, not important).

[7] The question is: "What are your main reasons for using social media."

When it comes to their perception of what they have learned in school related to citizenship, it seems that the school is dealing with most of these topics, especially the environment, with the exception of information about the legislative process.[8]

Of course, this universe is far from being representative of the state of citizenship education in the public secondary schools of Rio de Janeiro, but it may provide some glimpse of the situation and allow for the elaboration of some hypotheses about which topics should be better dealt with in those schools and which competences might be more emphasized.

That said, on the one hand, we believe that only major structural changes can resolve problems such as underpayment and overwork: on the other hand, we consider that small improvements are made through the development of improved citizenship education, especially in its social–emotional and behavioral aspects.

10.4 Citizenship Education as a Case Study for an Innovative Approach: Methodology and Path Followed

As a part of the project "Urban regimes and citizenship: a case study for an innovative approach" (funded by the Swiss National Research Foundation (SNSF), CNPq and FAPERJ) was set up and tested a module of citizenship education, set up through a sequence of steps.

An initial quantitative survey was conducted, defining two indices from 2010 to 2016, one capturing characteristics of an urban regime and the second focusing on citizenship indicators.

Admittedly, the setup of the indices presents some weak points in terms of some data comparability, and in the difficulty to compare Rio de Janeiro city with other peripheral municipalities at social, political, and economic levels, due to the relevant differences between Rio and its peripheries. Despite this bias, the indices have allowed for the grasp of relevant information.

The overall situation shows a general increase in both urban regime and citizenship indices. This is reflected in Fig. 10.1, where each dot, representing one of Rio de Janeiro's 19 Metropolitan Region (RJMR) municipalities, is placed according to their situation in 2010 and 2016. The grey line shows where dots would lay if there were no evolution between the two periods of time. In terms of citizenship, all the municipalities show an increase in intensity, albeit to different degrees.

While municipalities evolved more heterogeneously for the urban regime index, we still observe an overall increase (on average the 19 municipalities rose from 0.406

---

[8]The question is "In school, have you learned about: how elections work in Brazil; how laws are made?; how to protect the environment?; how to contribute to help solving problems of the community?; other countries' social and political problems?". To each of the topics of the question, they could answer yes or no.

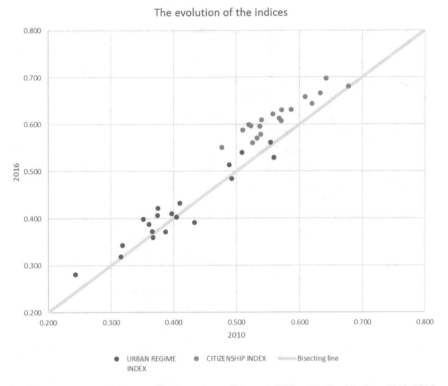

Fig. 10.1 The trajectory of both (democratic conditions and Urban Regime) indices 2010–2016. *Source* own elaboration

to 0.417) but in some cases, a decrease is observed. This is due to the democratic condition subindex (Fig. 10.1), the only one that shows a decline.

Despite the general upward trend, the evolution within municipalities is heterogeneous but the employment index and education conditions represent the worst indicators of the RJMR.

The different behavior of the two indices, within municipalities, is reflected in Fig. 10.2, showing the correlation between them. The linear trend line reflects the correlation between the two indices in 2010 (blue) and 2016 (orange). The correlation is clearly decreasing and nearly intangible in 2016.

The correlations show a positive course of 0.389 (statistically significant at 90%) for 2010, which means that for every percentage point of increase of the urban regime index, the citizenship index rises by 0.389 percentage points. However, for 2016, this correlation drops to 0.275 and became statistically not significant, showing a decrease in trend.

In terms of correlation between subindicators, we observe that the citizenship index is correlated to the democratic conditions (+0.676) and infrastructure (+0.457) subindicators for 2010, and only with the democratic conditions (+0.577) in 2016. This would mean that in order to increase the citizenship indicator, the democratic

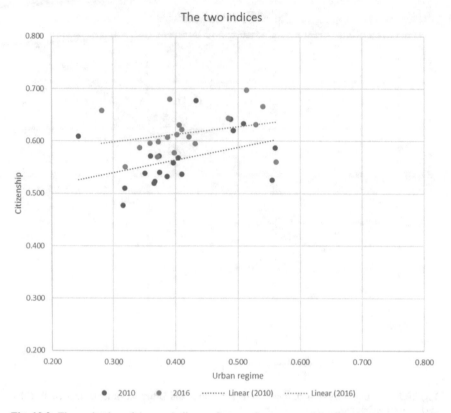

Fig. 10.2 The evaluation of the two indices trajectory. *Source* own elaboration

condition indicators could be concentrated on. On the other hand, the urban regime index is correlated with education (+0.564) and social vulnerability (+0.524) subindicators in 2010 and only with social vulnerability (+0.695) in 2016. Therefore, for a better outcome in terms of urban regime, one could intervene in the social vulnerability aspects.

The output of this quantitative work demonstrates that the link of the two indices and the comparison over time suggest that some elements seem particularly meaningful and crucial, regarding the main three areas where intervention might improve the frame of both indices, namely: (a) democratic conditions, (b) social vulnerability, and (c) education.

The connections found and the three areas emerged summarize the need for people to be enabled to develop their capability of thoughtful and responsible participation in political, economic, social, and cultural life. The results of the indices suggest that it depends on the development of two "core" themes:

- the first is that people learn more about citizenship by being active citizens, by participating (in a wider sense) in social and political life. This requires that education strategy start modeling the kind of society in which participation (as an enabler of democratic conditions) is encouraged by providing all with opportunities to take responsibility, increase awareness, and exercise choice;
- the second is that the development of capability for citizenship should be fostered in ways that aim to reduce social vulnerability by motivating and equipping people (in particular the young) to be active and responsible members of their communities, at local, national, and global levels.

Such results demonstrate that citizenship education is one of the most important emerging topics in the educational field (see UNESCO 2015; Torres 2015), since citizenship education is not currently an option for policy and decision-makers, but merely seems to be a way to avoid social and political exclusion, a prevention of harsh social conflicts. In other words, as citizenship is constitutive of rights and responsibility and since *"who can exercise and claim these rights is itself contestable, citizenship is practiced not only by exercising these rights but also by claiming them"* (Isin 2017b, p. 501).

A further step was the definition of delivery method. Given the context and objectives developed, the most appropriate approach to achieve the objectives seemed to combine PBL (Problem Based Learning) and debate. As a mode of collaborative learning based on problem analysis and resolution, PBL stands as an appropriate methodology, due to its flexibility and transversal applicability with secondary school students. It is an approach that leads to an appreciation of both the theoretical and practical life experience, based on promoting autonomy, in an organized manner, and the development of the ability to access necessary problem-solving resources.

According to some authors (Panlumlers and Wannapiroon 2015; De Graaff and Kolmos 2003), PBL is generated by the work process finalized to comprehend or solve a problem. PBL, which has evolved through the years, lends itself to different methodological interpretations and developments (Kolmos 2008).

The second approach combined with PBL, debate in the class, cultivates an active engagement of students, allowing participation of the whole class divided into groups (Kennedy 2009). In the vast options of debate, we selected a form of debate where the whole class is involved, assigning a topic and forming three groups. One is in favor and one against the topic, and the third listens to the two opposite positions and formulates a final decision, considering both positions and motivating their reasons. In the project we then adopted a form of "speed" debate, structured to be appropriately tested, optimizing time and outputs in term of interaction, ability to exploit the topic facets, and the possibility to reach a "conclusion" in one lesson units.

The benefits of using a debate as an instructional strategy also include mastery of the content and the development of critical thinking skills, negotiations, finding an acceptable final deliberation, participating in a process, stimulating the attention, and fostering the appropriation of learning in students and oral communication skills (Kennedy 2007; Zare and Othman 2015).

Based on the aforementioned conceptions of citizenship, taking into account Rio de Janeiro's specificities in terms of democratic conditions and social vulnerability raised from the quantitative initial findings and considering the potential of PBL and speed debate teaching methods as tools to develop cognitive and behavioral and socioemotional abilities essential to an active citizen, a module was elaborated in order to give secondary school teachers guidelines and activity suggestions meant to be helpful in their everyday efforts to deal with the complex realm of citizenship education.

The citizenship education module is divided into: learning objectives, contents, activity suggestions, bibliography of support.

Though a vast literature was consulted, there are four important references worth mentioning here for the elaboration of the module: global citizenship education principles (UNESCO 2015); behavioral dispositions adequate to a democratic life (Benevides de Mesquita 1996); Isin's works on enabling an international and performative citizenship; Rawls' conception of the political basis constituting a liberal and participatory society.

In other words, the module aims to combine the development of the ability to gain knowledge in order to make better choices, to better judge, to ingrain habits of tolerance in the face of differences as well as to learn active cooperation and the subordination of personal or group interests to the general interest and the common good. This is especially needed in a context of growing disbelief in politics, the pervasiveness of fake news in the social media and the crisis of traditional media, growing levels of urban violence, and a strengthening of ultraconservative interpretations of the roots of our problems (Messenberg 2017). Such is the current scenario in most big cities in Brazil, especially in Rio de Janeiro.

Although the test takes place in only one school, other teachers could have access to a valid and theoretically based guide to implement some appropriate activities in terms of citizenship education.

10.5   The Citizenship Education Module as a Result Tested in School

Given the theoretical framework illustrated in the above section on the fundamental contributions that citizenship education can deliver to society, the citizenship education module finally completed and tested (after a process of elaboration and content adjustments, also including main stakeholders), stands on three legs. The module has been reprocessed and amended, in order to deliver a simple and clear tool to teachers. It has been studied to equip them with a structure of a module immediately exploitable, with the necessary completion of two instructional workshops and a minimal theoretical background study based on specific literature on citizenship and citizenship education.

The citizenship education module is then constituted by a method of delivery based on:

(a) two workshops, with the teachers adopting the module, delivered by project experts;
(b) a basic theoretical literature to be studied indicated by project experts. This literature can be adapted in context (e.g., for Rio de Janeiro some articles on the specific citizenship curriculum at school are given) but is focused on the theories of citizenship and on understanding the basic principle of teaching them;
(c) a document describing the citizenship education module for teachers.[9]

This last document consists of three parts, each structured as follows:

- brief introduction of the aims;
- learning objectives;
- contents (a list of learning objectives);
- example of implementation, in the form of debate and/or PBL.

The three parts of the citizenship education module for teachers are settled as follows, including a brief overview of a basic conceptual explanation of the contents behind each one:

- **Part 1**: Citizenship in and beyond the nation–state in a context of interdependence (globalization).
- **Part 2**: The contexts of citizenship: institutions, intergovernmental organizations, governance, accountability, and citizen participation.
- **Part 3**: The city as laboratory and pacer of citizenship.

Some parts of this module were tested in a public secondary school in Rio de Janeiro. The research team followed the test in two groups, both with the same teacher. In one of the groups, the teacher discussed with the student the struggle for the right to choose representatives through elections. The debate, as proposed in the module, was about whether voting should be mandatory or not. In the other group, followed by the research team, the teacher discussed democracy and representation and used the PBL technique in an elaboration of "class constitution."

In general, the experience was positive. In the case of the debate, debating teams covered the main arguments for and against a mandatory vote in Brazil. Students were also quite respectful to each other, paying close attention to their time and not interrupting the opposing team. In the test involving PBL, students were participative and talkative. Important issues emerged, both about the form and the content of a legislation/regulation. At the end of the process, they were much more enlightened about the substance of legislation in general.

Questioned about the activities format, they were quite enthusiastic about a class format different from the lecture. They valued the experience of being more active

---

[9]The final version of the citizenship education module delivered to teachers, in the Portuguese language can be accessed here: http://www.supsi.ch/deass/ricerca/banca-dati-progetti/in-evidenza/Urban-regimes-and-citizenship--a-case-study-for-an-innovative-approach.html.

during class and having the opportunity to have a more open space to talk to each other. Some students also suggested thoughtful adjustments to the dynamic of the debate.

10.6 Conclusion

There are many sure benefits to citizenship education. Who could deny the importance of a more peaceful, more just, safe, and sustainable context in which to live? In addition to these obvious long-term benefits, there are also immediate benefits. Studying global problems and the various strategies to address them on an individual level can generate a renewed sense of participation, membership, and optimism. Practicing citizenship, whether through personal changes, service learning, grassroots organizing, or other activities promoting individual motivation, can provide meaning to the school curriculum. Teachers and students can all see that they make an impact on society, starting from their neighborhood and their city, beyond the individualistic goal of reaching a good position in society or just obtaining a job.

The citizenship education module set up and tested in the Rio de Janeiro context, we can conclude, aims both to strengthen individual motivation, and to stimulate critical thinking and participation in the social process. This aspect seems crucial to kick start an individual (and then collective) awareness, able to balance, in the long term , an invasive "*financial pantheism*" (Bignami 2017, p. 133). Such a concept goes beyond financial colonization, to the economy. It indicates the occupation of institutions governing the states from the interdependence of economic and financial mechanisms. The accent here is on individual.

At the same time, we tested citizenship education by taking into account not only disputes surrounding the privilege of the nation–state as the appropriate scale of the political community, but also the acceptance of liberal institutions themselves and their underlying values, considered as threats by some conservative and fundamentalist groups (Kymlicka 2003, p. 48). Since education historically has in itself, as a part of its aim, the consolidation of a national identity and membership among citizens, today citizenship education is also challenged by supranational forums, such as the European Union, and from universalistic claims in general, such as global citizenship, but also from communitarian and local interests, and from the individual's expectation of respect for personality and preferences. Here the accent, therefore, is on collectivity.

In terms of perspective, inspiration from UNESCO's Global Citizenship Education principles (the three domains of learning cognitive, socioemotional, and behavioral) seems fruitful to maintain a certain linearity and homogeneity between individual and collective learning outcomes. This inspiration takes into account some bias of such principles; it seems indeed opportune to reflect upon this approach to be used in scaling up citizenship education, since it has influence at the political (and polity) level for neoliberal, radical, or transformative implications (VanderDussen Toukan 2018) linked to values disseminated in different contexts.

At the policy-making level, this ambitious goal requires a shared effort with educational institutions and policy-makers convinced that the school curriculum (especially for secondary school) needs a turning point on the concept of citizenship; at the implementation level, on the one hand, an effective strategy to strengthen the experimentation of active-learning methodologies, such as debate and PBL, in perspective supported by ICT as facilitator for delivery and attractiveness of the subject for both teachers and students. Though these methodologies are not new, they are still not widespread in secondary schools, which rely mostly on lectures and evaluations that concentrate on cognitive aspects, nor are they adopted for citizenship education. At the same time, teachers, equipped with both competence and citizenship sensitivity, appears to be the main route to follow.

References

Alsayyad N, Roy A (2006) Medieval modernity: on citizenship and urbanism in a global era. Space Polity 10(1):1–20

Barber BR (2008) Consumed: how markets corrupt children, infantilize adults, and swallow citizens whole. W.W. Norton, New York

Benevides de Mesquita MV (1996) Educação para a democracia. Lua Nova 38:223–237. http://www.scielo.br/scielo.php?script=sci_arttext&pid=S0102-64451996000200011&lng=en&nrm=iso

Bignami F (2017) Going intercultural as a generative framework of a respondent citizenship. In: Onorati MG, Bignami F, Bednarz F (eds) Intercultural praxis for ethical action. Reflexive education and participatory citizenship for a respondent sociality. EME publications, Louvain, Belgium

Bignami F, D'Angelo V, Bednarz F (eds) (2016) New educational itineraries and perspectives for care professionals—The innovative caregivers' Training model as example of strenghtening competences, networking and participation. Scholars' Press, Saarbrücken

Corak M (2016) 'Inequality is the root of social evil', or maybe not? Two stories about inequality and public policy. Canadian Public Policy 42(4):367–414

Dagger R (1997) Civic virtues. Rights, citizenship, and republican liberalism. Oxford University Press, Oxford

De Graaff E, Kolmos A (2003) Characteristics of problem-based learning. Int J Eng Educ 19(5):657–662

Elster J (1989) The cement of society. a survey of social order. Cambridge University Press

Galston WA (1991) Liberal purposes. Cambridge University Press

Galston WA (2005) The practice of liberal pluralism. Cambridge University Press, Cambridge

Geboers E, Geijsel F, Admiraal W, ten Dam G (2013) Review of the effects of citizenship education. Educ Res Rev 9(2013):158–173

Holston J (2008) Insurgent citizenship: disjunctions of democracy and modernity in Brazil. Princeton University Press, Princeton

Isin EF (2017a) Enacting international citizenship. In: Basaran T, Bigo D, Guittet E-P, Walker RBJ (eds) International political sociology: transversal lines. Routledge, London

Isin EF (2017b) Performative citizenship. In: Shachar A, Bauböck R, Bloemraad I, Vink M (eds) The Oxford handbook of citizenship. Oxford University Press, Oxford, pp 500–523

Kennedy R (2007) In-class debates: fertile ground for active learning and the cultivation of critical thinking and oral communication skills. Int J Teach Learn High Educ 19(2):183–190. http://www.isetl.org/ijtlhe/

Kennedy R (2009) The power of in-class debates. Act Learn High Educ 10(3):1–12. https://doi.org/10.1177/1469787409343186

Kolmos A (2008) Problem-based and project-based learning. In: Skovsmose O, Christensen P, Christensen OR (eds) University science and mathematics education in transition. Springer, London, pp 261–282

Koslinski M, de Queiroz Ribeiro LC (2017) Segregation and educational inequalities. In: de Queiroz Ribeiro LC (ed) Urban transformations in Rio de Janeiro - development, segregation, and governance. Springer, pp. 165–190

Koslinski M, de Queiroz Ribeiro LC (2014) Urban frontiers and educational opportunities: the case of Rio de Janeiro. In: de Queiroz Ribeiro LC (ed) The metropolis of Rio de Janeiro. A space in transition. Letra Capital Editora, Rio de Janeiro

Kymlicka W (2003) Two dilemmas of citizenship education in pluralist society. In: Lockyer A, Crick B, Annette J (eds) Education for democratic citizenship. Ashgate Publishing Limited, Hants, pp 47–63

Lin AR, Fahey Lawrence J, Snow CE (2015) Teaching urban youth about controversial issues: pathways to becoming active and informed citizens. Citizsh Soc Econ Educ 14(2):103–119

Marshall TH (1950) Citizenship and social class and other essays. Cambridge University Press, Cambridge

Maynor JW (2003) Republicanism in the modern world. Polity Press, Cambridge

McIntosh H, Youniss J (2010) Toward a political theory of political socialization of youth. In: Sherrod L, Torney-Purta J, Flanagan CA (eds) Handbook of research on civic engagement in youth. John Wiley & Sons, Hoboken, NJ, pp 23–41

Menezes I (2003) Participation experiences and civic concepts, attitudes and engagement: implications for citizenship education projects. Europ Educ Res J 2(3):430–445

Messenberg D (2017) A direita que saiu do armário: a cosmovisão dos formadores de opinião dos manifestantes de direita brasileiros. Revista Sociedade e Estado 32(3):621–647. https://doi.org/10.1590/s0102-69922017.3203004

Mezzadra S, Neilson B (2012) Between inclusion and exclusion: on the topology of global space and borders. Theory Cult Soc 29(4/5):58–75

Nguyen J (2018) Identity, rights and surveillance in an era of transforming citizenship. Citizsh Stud 22(1):86–93. https://doi.org/10.1080/13621025.2017.1406456

Panlumlers K, Wannapiroon P (2015) Design of cooperative problem-based learning activities to enhance cooperation skill in online environment. Procedia Soc Behav Sci 174:2184–2190

Pinkett RD (2000) Bridging the digital divide: socio-cultural constructionism and an asset-based approach to community technology and community building. Paper presented at the 81st Annual Meeting of the American Educational Research Association (AERA). April, New Orleans, USA

Rawls J (2005) Political liberalism. Expanded edition. Columbia University Press, New York

Sassen S (2002) The repositioning of citizenship: emergent subjects and spaces for politics. Berkeley J Sociol 46

Sassen S (2005) The repositioning of citizenship and alienage: emergent subjects and spaces for politics. Globalizations 2:79–94

Torres CA (2015) Global citizenship and global universities: the age of global interdependence and cosmopolitanism. Europ J Educ 50(3):262–279

UNESCO (2015) Global citizenship education - topics and learning objectives. UNESCO- United Nations Educational, Scientific and Cultural Organization, Paris

VanderDussen Toukan E (2018) Educating citizens of 'the global': mapping textual constructs of UNESCO's global citizenship education 2012–2015. Educ Citizsh Soc Just 13(1):51–64. https://doi.org/10.1177/1746197917700909

Walzer M (1989) Citizenship. In: Ball T, Farr J, Hanson RC (eds) Political innovation and conceptual change. Cambridge University Press, Cambridge

Warleigh A (2006) Learning from Europe? EU studies and the re-thinking of "International relations". Europ J Int Relat 12(1):31–51

WEF (2018) The global risks report 2018–13th Edition. World Economic Forum, Geneva

Westheimer J, Kahne J (2004) What kind of citizen? The politics of educating for democracy. Am Educ Res J 41(2):237–269

Zare P, Othman M (2015) Students' perceptions toward using classroom debate to develop critical thinking and oral communication ability. Asian Soc Sci - Published by Can Center Sci Educ 11(9):158–170

Index

Printed in the United States
by Baker & Taylor Publisher Services